Communication Theory

Communication Theory

Media, Technology, Society

David Holmes

SAGE Publications
London ● Thousand Oaks ● New Delhi

First published 2005

SAGE Publications Ltd
1 Oliver's Yard
55 City Road
London EC1Y 1SP

SAGE Publications Inc.
2455 Teller Road
Thousand Oaks, California 91320

SAGE Publications India Pvt Ltd
B-42, Panchsheel Enclave
Post Box 4109
New Delhi 110 017

British Library Cataloguing in Publication data

A catalogue record for this book is available
from the British Library

ISBN 0-7619-7069-X
ISBN 0-7619-7070-3 (pbk)

Library of Congress Control Number available

Typeset by C&M Digitals (P) Ltd, Chennai, India
Printed in Great Britain by TJ International, Padstow, Cornwall

For Elena

Erratum

p.199, paragraph 3

For Mike Featherstone (1998: 910), in the modern period this means that urban spaces where the occupants of different residential areas could meet face-to-face, engage in casual encounters, accost and challenge one another, talk, quarrel, argue or agree, lifting their private problems to the level of public issues and making public issues into matters of private concern - those 'private/public' *agora* of Cornelius Castoriadis are fast shrinking in size and number.

This should read:

For Zygmunt Bauman, in the modern period this means that 'urban spaces where the occupants of different residential areas could meet face-to-face, engage in casual encounters, accost and challenge one another, talk, quarrel, argue or agree, lifting their private problems to the level of public issues and making public issues into matters of private concern - those "private/public" *agoras* of Cornelius Castoriadis's are fast shrinking in size and number' (1998: 21).

This quote comes from:

Bauman, Z. (1998) *Globalization: The Human Consequences*. Columbia, NY: Columbia University Press.

Other corrections to *Communication Theory* can be found on the Sage Publications website:
www.sagepub.co.uk/holmes

CONTENTS

List of Tables and Figures   ix
Preface   x
Acknowledgements   xv

1 **Introduction – A Second Media Age?**   **1**

   Communication in cybercultures   3
   The overstatement of linguistic perspectives on media   4
   The first and second media age – the historical distinction   7
   Broadcast mediums and network mediums – problems
   with the historical typology   11
   Interaction versus integration   15

2 **Theories of Broadcast Media**   **20**

   The media as an extended form of the social – the rise
   of 'mass media'   21
   Mass media as a culture industry – from critical theory to
   cultural studies   23
   The media as an apparatus of ideology   25
   Ideology as a structure of broadcast – Althusser   29
   The society of the spectacle – Debord, Boorstin and Foucault   31
   Mass media as the dominant form of access to social
   reality – Baudrillard   36
   The medium is the message – McLuhan, Innis and Meyrowitz   38

3 **Theories of Cybersociety**   **44**

   Cyberspace   44
   Theories   50
   Social implications   72

4 **The Interrelation between Broadcast and
   Network Communication**   **83**

   The first and second media age as mutually constitutive   83
   Broadcast and network interactivity as forms of
   communicative solidarity   86

Understanding network communication in the
context of broadcast communication 97
Understanding broadcast communication in the
context of network communication 101
Audiences without texts 111
The return of medium theory 113
Recasting broadcast in terms of medium theory 118

5 **Interaction versus Integration** **122**

Transmission versus ritual views of communication 122
Types of interaction 135
The problem with 'mediation' 138
Medium theory and individuality 140
Reciprocity without interaction – broadcast 144
Interaction without reciprocity – the Internet 149
The levels of integration argument 151

6 **Telecommunity** **167**

Rethinking community 167
Classical theories of community 168
The 'end of the social' and the new discourse of community 171
Globalization and social context 173
The rise of global communities of practice 174
Sociality with mediums/sociality with objects 177
Post-social society and the generational divide 186
Network communities 188
Broadcast communities 206
Telecommunity 221

References 226

Index 244

LIST OF TABLES AND FIGURES

Tables

1.1 The historical distinction between the first and
    second media age   10

3.1 Digitalization as the basis of convergence, wider
    bandwidth and multi-media (the ability to combine
    image, sound and text)   66
3.2 Features and types of hot and cool mediums   71

4.1 The broadcast event   105
4.2 Medium theory as applied to network and
    (retrospectively) to broadcast communication   119

5.1 Transmission and ritual perspectives compared   135
5.2 John B. Thompson's instrumental/mediation paradigm   137
5.3 Broadcast and network as forms of communicative
    integration   149

Figures

3.1 Transmission model: high integration/low reciprocity   53

5.1 Ritual model: high integration/high reciprocity   147

PREFACE

A theory of communication must be developed in the realm of abstraction. Given that physics has taken this step in the theory of relativity and quantum mechanics, abstraction should not be in itself an objection.

N. Luhmann, *Art as a Social System*, trans. Eva M. Knodt, Stanford: Stanford University Press, 2000, p. 12

What follows is an interdisciplinary communication theory book which sets out the implications of new communications technologies for media studies and the sociology of communication.

The cluster of texts which came out over the last decade dealing with computer-mediated communication (CMC), virtual reality and cyberspace has significantly established new theoretical domains of research which have been accepted across a range of disciplines. The current book proposes to integrate this literature in outline and summary form into the corpus of communication studies. In doing so it explores the relationship between media, technology and society. How do media, in their various forms, extend the social, reproduce the social, or substitute for other aspects of social life?

Most books dealing with communication and media studies invariably address traditional concerns of content, representation, semiotics and ideology. Whilst including an appreciation of these approaches, the current book makes a contribution to theoretical analysis of media and communications by charting how the emergence of new post-broadcast and interactive forms of communication has provided additional domains of study for communication theory, renovated the older domain of broadcast, and suggested fresh ways of studying these older media.

In doing so, this book advances a critique of the 'second media age' thesis, which, I argue, has become something of an orthodoxy in much recent literature. It rejects the historical proposition that a second media age of new media, exemplified by the Internet, has overtaken or converged with an older age of broadcast media. Yet at the same time, the value of analytically distinguishing between the most significant architecture that is attributed to the first media age – broadcast – and that which is attributed to the second media age – interactive networks – is upheld. The basic dualism between broadcast and interactivity structures the main themes of the book. To the extent that individuals in media societies experience changes in the means of communication as a 'second media age', we are compelled to re-examine the postulated 'first media age' in terms of medium or network form rather than simply content or 'text'. The sense in which this distinction is made should not be confused with questions of form versus the content of narrative, where content is what a text says, and the form is how it says it. Rather, a non-textual distinction is being made here. In doing so, a sociological appreciation of broadcast can be arrived at rather than a media studies or cultural studies perspective, which is invariably grounded exclusively in either behaviourist or linguistically centred approaches to analysis. However, insofar as this book is 'sociological', sociology is not being opposed to communication and media studies; on the contrary, a central argument of the book is that emergence of new communication environments has more or less forced traditional media and communication studies to be sociological. For this reason the current volume is very interdisciplinary (between communication, media and sociology), but this has less to do with the perspective adopted than with changes in how media are experienced.

These recent changes in media infrastructure have necessitated a shift in the order in which communication theory is treated. For example, information theory, which often prefigures semiotic analysis of media, is introduced in the current textbook as instructive for the second media age, where it more appropriately belongs with analyses of the Internet. In fact, in seeing just how relevant information theory is to CMC rather than broadcast, it is surprising how significantly it came to figure in studies of broadcast in the first place. At the same time, the book tries to incorporate most of the traditions of twentieth-century communication theory in order to locate their relevance to studying the sociological complexities of contemporary convergent communications.

Through this argument the distinction between medium and content, media and messages, is persistently returned to. On the scaffold of these distinctions the book also presents a central argument about the difference between communicative *interaction* and *integration*. With the aid of recently emerging 'ritual' models of communication it is possible to understand how the technical modes of association manifested in broadcast and interactive communication networks are constitutive of their own modes of integration. Thus it is possible to identify media-constituted communities in broadcast communities and so-called 'virtual communities', which is to argue that such networks do not so much 'mediate' *interaction,* as facilitate modes or levels of *integration* to which correspond specific qualities of attachment and association. It is also to argue that media-constituted communities aren't merely a continuation of older face-to-face or geographic communities by technical means (the mediation argument) but are rather constitutive of their own properties and dynamics. Of course, such 'levels' of integration are not isolated but co-exist, in ways which are outlined in successive chapters (particularly Chapters 4 and 5). A third major theme that is explored is the urban and economic context of media-constituted communities, the way in which dependence on technical-communicative systems facilitates expanded commodification and rationalization of cultural life: spheres which could never have been so influenced before the emergence of these systems.

It is not only the second media age text which is to be reappraised in developing the book's themes but also some classical texts on the sociological dynamics of broadcast as well as key readers pertaining to frameworks of 'media studies'. Where this book differs from 'media studies' texts is in integrating the significance of 'cybersociety' into the general corpus of communication theory. It does so by way of a critique of the second media age orthodoxy which imagines a new era that is derived from yet another progress-driven 'communications revolution'. At the same time, the discourses of 'telecommunications convergence' are critically assessed for overstating a technologically reductive distinction between 'broadcast' and 'interactivity' in order that they can be portrayed as undergoing 'convergence', again at a solely technological level.

To turn to the chapter composition of the book: the introduction establishes the rationale guiding the organization of the book: the contrast between broadcast and network forms of communication. The predominance of semiotic accounts of media is criticized as unwarranted, distracting attention from the techno-social dimensions of media environments. At the same time, a linear model of progression from a first to a second media age is found to be too simplistic to address the complexity of contemporary media formations. The linear model is premised largely on an interaction approach to media culture, which in this chapter is counterposed to the more fruitful analyses that are made possible by 'integration' models. A variant of the linear second media age perspective is the 'convergence' thesis, which presupposes two media forms (of broadcast and interactivity) not historically, but technologically. These themes, of first versus second media age, of a multiplicity of form versus content, of 'convergence' as a product of medium dichotomization, of interaction versus integration, are announced as guiding the development of the whole volume.

Chapters 2 and 3 are stand-alone expositions of theories of 'broadcast communication' and 'network communication', respectively. These chapters introduce key theoretical perspectives that are relevant to understanding broadcast and network communication. In addition, an historical and empirical discussion of broadcast in the context of urbanization and the rise of industrial society is presented, whilst in Chapter 3 the major innovations which underlie the second media age thesis are considered. Chapter 2 reproduces much of the 'classical' literature on media (e.g. theories of ideology) whilst also recasting it within the macro-framework of the techno-social medium approach (e.g. Althusser's often difficult theory of 'interpellation' and 'ideology-in-general' is re-explained as an effect of the structure of broadcast. Chapter 3 attempts to formalize the still very young perspectives on cybersociety and proposes to give them a sense of definition as a way of ordering the current burgeoning literature. In doing so, it identifies a 'second media age' perspective, a CMC perspective, convergence perspectives and the reclamation of older perspectives (McLuhan, Baudrillard) whose relevance to cyberculture is arguably greater than it is to media culture.

Chapter 4 considers the interrelation between broadcast and network mediums[1], and argues that they are quite distinct in their social implications but are also parasitic on each other. In this light, what is called 'convergence' is really an outcome, rather than a cause, of such parasitism, a consequence which is mistakenly seen to be only working at the level of technical causation, or predestined historical *telos*. But this distinctively broader meaning of convergence can only be arrived at if correspondingly broader meanings of network and broadcast are deployed, to spheres not confined to media and communications. In the context of such criticism, media technologies, whether they be broadcast or interactive, increasingly reveal themselves as *urban* technologies, which are constantly converging with the logics internal to other urban technologies (the shopping mall, the freeway). For example, the argument that virtual communities restore the loss of community that is said to result from the one-dimensionality of the culture industry does not contrast virtual and 'physical' communities, which can be done by looking at the dialectic between media culture and urban culture. Raymond Williams' under-regarded concept of 'mobile privatization' is explored as a departure point for the way in which media extend social relations on the basis of private spatial logics.

Finally, the economic complementarity of broadcast and network mediums is established. Life on the screen is one in which individuals are, if they so choose, able to live a culture of communication without the spectacle and advertising fetishes of broadcast. However, in an abstract world of communicative association this new mode of 'communication as culture' itself provides a market for communication products, both hardware and software, that is growing on a scale which is rapidly catching up with the political economy of broadcast.

Chapter 5, 'Interaction versus Integration', critiques various models of interaction (instrumental views of communication, transmission views, 'mediation' views) as not being able to adequately address the socializing and socially constituting qualities of various media and communication mediums. In doing so it turns its attention toward the promising body of theory which can be gathered under the heading of 'ritual communication'. This comprises works such as James Carey's *Communication as Culture* and is informed by anthropological perspectives and New Media theory. An argument is made for the need to develop an understanding of 'levels' of ritual communication: face-to-face, mediated and technically extended. The advance that John B. Thompson makes in this regard in *The Media and Modernity* is a useful stepping stone, but one that is based on interaction rather than 'integration'. Integration formulations (Meyrowitz, Calhoun, Giddens) are then explored in order to demonstrate the shortfalls of the interaction model as well as to sketch a model which can begin to attend to the complexity of both broadcast and network forms of communication processes.

Chapter 6, on telecommunity, appraises the significance of the concept of community in media culture in two ways. Firstly, how do 'communities' arise that are said to be constituted entirely by technical mediums? Secondly, why is it only recently, after over a hundred years, that there has been a radical renewal of thinking of community? With regard to the first question, the idea of a virtual community is explored, but in relation to the much neglected idea of broadcast communities, which, if anything, offer more powerful forms of integration than do their cyberspace counterparts. Whereas in broadcast communities there is little or no interaction with others in embodied or quasi-embodied form, there is a high concentration of identification and the constitution of community by way of extended charismatic affect. Thus, both kinds of community can be characterized as virtual in the way in which they privilege relations with media and mediated association.

In its emphasis on the priority of techno-social mediums over content, the volume draws on the recent wave of publications that have dealt with the Internet and communication theory. At the same time it attempts to chart the relationship between traditional and new media without exaggerating the impact of the latter. Not only does broadcast remain central to modern media culture, but it makes possible, in co-dependent ways, the social conditions which underpin cyberculture, from its first steps to its last.

Note

1 Whilst the term 'media' might normally be considered the plural of medium, in this book I make the distinction between media and mediums which is not restricted to a singular/plural distinction. In using 'mediums' I am trying to retain a strong sense of media as environments, rather than as either 'technologies' or institutions. Denoting 'mediums' as 'media environments' or 'media architectures' facilitates insights drawn from medium theory which cannot be served by the term 'media'.

ACKNOWLEDGEMENTS

The analysis presented in this book has emerged from almost ten years of teaching and researching sociology of media. I have been fortunate to present my research across sociology and communications forums, and have benefitted from the challenges of organizing my ideas for classrooms of inquiring minds. I am grateful to the Humanities Research Program at the University of New South Wales for providing assistance in the middle phase of writing, as well as colleagues both there and at Monash for ongoing conversation, encouragement and thoughtfulness. In particular, I would like to thank Paul Jones and Ned Rossiter for rewarding conversations around themes and figures central to the book. For assistance with some research tasks, thanks are due to Aaron Cross and Olivia Harvey. With production I would like to Fabienne Pedroletti at Sage Publications and in Melbourne, Andrew Padgett, who has been invaluable with the final productive phases. Finally, my gratitude to my partner Vasilka Pateras for her patience and love, and our daughter Elena, for being an aspiration.

ONE

INTRODUCTION – A SECOND MEDIA AGE?

> In the last few years … widespread talk of 'cyberspace' has brought new attention to the idea that media research should focus less on the messages and more on communication technologies as types of social environments. (Meyrowitz, 1999: 51)

In an essay, 'Learning the Electronic Life', written just before the 'widespread talk of cyberspace' that accompanied the so-called 'Internet Revolution' of the 1990s, James Schwoch and Mimi White (1992) set about to describe a typical day's activity for their American family – from waking up, to putting in hours as teachers in the education sector, to trying to relax in the evening.

At first light they relate how they are woken by the baby monitor which links their room to their son's. Next thing they are heating up the rice cereal in a microwave. While their boy is in the playpen, James and Mimi commence some exercise in front of the TV with remote control handy.

> Out of the house and, if not a walk-to-work day, into the car, lowering the garage door with the automatic opener as we drive away on errands. Stop at the bank – or rather, the nearest automatic teller machine to get some cash for groceries and shopping (done with cash, checks, and credit cards, with access to the first electronically verified by a local computer network, the latter two verified at point of purchase by a national computer network) – and upon returning home, check the phone machine before going off to the office or upstairs to the study to work on the computer. A typical work day can include not only personally interacting with students and colleagues, but also interfacing with long distance telephone calls, photocopies, print-outs, hard drives, programs, modems, electronic mail, floppies, audio and video tape, and once in a while a fax. If we do not work into the evening, a typical night may well include (along with returning phone calls) radio listening, recorded music (albums, tapes or compact discs), broadcast television, cable television, or videocassettes. The most probable result, of course, is some combination of the above choices, with too many TV nights degenerating into an uninspired channel-hopping via remote from the comfort of the couch. In the background the baby monitor provides the sound of sleeping baby, a sound that accompanies us into bed each evening. The cycle, with a slight degree of variation, begins anew the next day. (Schwoch and White, 1992: 101–2)

Schwoch and White describe these interactions as 'an unremarkable series of events' about which 'few stop to marvel at how quickly and unthinkingly certain aspects of technology – telecommunications based on the electromagnetic spectrum and various wire-based telecommunications networks such as the telephone – become part of our everyday experiences' (102). Their very prosaicness, they argue, is what makes them so important and powerful, because it is in our interface with these technologies, the human–technical interface, that an entire pedagogy of technical competence is fostered, a pedagogy which becomes almost buried in the thousands of discrete habits and routines that both help us, connect us and imprison us in the information society.[1]

People who live in information societies not only encounter and 'use' information and communication technologies; rather, increasingly, their modes of action are enframed by these technologies. They are not so much tools as environments. Since Schwoch and White published their essay, over a silicon century (seven years) has passed, in which time a range of interactive communication technologies have come become meaningful in our daily life. We could add to their scenario the emergence of digital, optic-fibre and packet-switching technologies which have made the Internet possible, and the normalization of satellite-based communications and information devices like satellite phones and global positioning systems (see Dizzard, 2000). More often than not, we are not even aware of the extent to which these technical systems precondition the simplest of activities – an ignorance which was aptly epitomized by the trillion-dollar anxiety over the millennium bug, the dreaded Y2K.[2]

But this lack of awareness does not signal that we have become 'overloaded' with information, images or technology, as subscribers to the 'saturation' thesis suggest.[3] Media saturation tends to encourage a view of some order of unmediated experience, which is menaced by impersonal scales of instrusive media. In this book, we will see that, in fact, attachment to media can be very personal and as meaningful as embodied relationships, and that appreciating the strength of these attachments requires a broadening of the concept of 'cyberspace'.

The exponential explosion in webs of CITs (communication and information technologies) has, at a phenomenological level, shifted the orientation many of us have to 'objects' to an extent that can change our sense of otherness.[4] As face-to-face relations are replaced by 'interface' with technological 'terminals' of communication, electronic devices acquire a life of their own. Outside our own bodies the world fills with objects that are also animated, an animation which might compete with the human – as suggested by Sherry Turkle's notion of the computer screen as a 'second self' (Turkle, 1984). Whilst the non-human might be competing with the human, individuals themselves increasingly find that they are part of contexts in which they are 'objectualized'.[5] Studies that have been conducted on these phenomena show high degrees of attachment to media and communication technologies, whether this be people's need to have a television on in the background even if they aren't actually watching it, the near desperation that many Internet users have in downloading their email, or individuals who find security in having a mobile phone even if they use it only seldom.

But of course, behind our surface contact with this system of objects are definite social relationships, relationships which new communication and information technologies enable to be *extended* in time and space (see Sharp, 1993). At the same time, however, the particular way in which they are extended can also be considered a relationship itself, which is capable of acquiring an independence from the function of extending 'pre-technological' or pre-virtual relationships, even if they somehow might take different kinds of reference from these relationships.

What this book proposes is that these electronically extended relationships are constitutive of their own dynamics, dynamics which can be studied beyond the bewildering array of object technologies which, in their very visibility, render the social relation largely invisible.

In particular, the social dynamics that will be analysed on the basis that they *can* be analysed as part of this technologically extended sphere of social integration are broadcast integration and network integration. By the end of this volume, I aim to show that these kinds of integration are ontologically distinct – that is, distinct in external reality, not just theoretically distinct – whilst at the same time mutually constitutive.

Communication in cybercultures

The technologically constituted urban setting which Schwoch and White describe is increasingly typical of contexts of everyday life which preside in the processes of modern communication. Communication does not happen in a vacuum, nor does it happen in homogeneous contexts or simply by dint of the features of a natural language, but in architectural, urban, technically and socially shaped ways.

This book explores the interrelation between these contexts and the character of a range of communication events. It is about the contexts of communication in so-called 'information' societies as well as the kinds of connection that these contexts and the communications themselves make possible. The urban and micro-urban realities that can be described in the everyday experiences of James and Mimi are integral to the understanding of contemporary communication processes. Is there a relationship between the increase in the use of CITs and the increase in the number of people living alone in America, Australia and Britain? Is there a logic which links the privatization of public space like shopping malls and the dependence on broadcast and network mediums?

In the last ten years, the convergence between technologies of urban life and new communications technologies has been remarkable. It has even led some commentators to argue that the privatizing concentration of so many context-worlds, be they electronic, architectural or automobile-derived, is what really amounts to 'cyberspace'. This convergence is perhaps nowhere more powerfully represented than it is by the Internet, which is itself a network as well as a model for 'cyberspace' relations.[6]

It was in the final decade of the twentieth century that the emergence of global interactive technologies, exemplified by the Internet, in the everyday sphere of advanced capitalist nations dramatically transformed the nature and scope of communication mediums. These transformations heralded the declaration of a 'second media age', which is seen as a departure from the dominance of broadcast forms of media such as newspapers, radio and television. Significantly, the heralding of a second media age is almost exclusively based on the rise of interactive media, most especially the Internet, rather than the decline of broadcast television. Empirically, some have pointed out how certain *technological forms* of mass broadcast have waned or fragmented in favour of 'market-specific communication' (see Marc, 2000), although this is seldom linked to the rise of extended interactive communication. Rather, what is significant for the second media age exponents is the rapid take-up of interactive forms of communication. Whether this take-up warrants the appellation of a second media age, which can so neatly signal the demise of a 'first media age', is contested in this book. Certainly, the second media age thesis points to and contains insights about definite changes in the media landscapes of nations and regions with high media density. But the conjunctive as much as the disjunctive relationships between old and new media are very important.

Nevertheless, the arrival of *what is described* as the 'second media age' has two important consequences: one practical and the other theoretical. The extent and complexity of these practical consequences, which this book outlines, concern the implications which 'the second media age' has for contemporary social integration. The theoretical consequence of the second media age is that it has necessitated a radical revision of the sociological significance of broadcast media as addressed by traditions of media studies.

The overstatement of linguistic perspectives on media

Under the influence of cultural studies, European traditions in media studies have, since the 1970s, typically focused on questions of content and representation rather than 'form' or 'medium'. This is perhaps itself a reaction to the preoccupation which 'process' models developed in the United States had with 'media effects' and behavioural epistemologies.[7]

Analysing media content – the employment of perspectives on language, beginning with Marxist conceptualizations of ideology, followed by the influence of 'semiotics', 'deconstruction' and 'New Criticism' – was conceived as a matter of studying the meaning of texts and discourse and the way in which the 'mass' media influence cultural values and individual consciousness. Throughout the 1970s and 1980s, differences between these approaches to studying texts were debated around the problem of social reproduction and how dominant discourses of a 'dominant ideology' were related to broader social form.[8] Under the umbrella of the linguistic paradigm, media studies has also concerned itself with 'media' over 'medium' – with the textuality of writing, still and moving images, music and speech – more than with the institutionalized adoption of these media in broadcast and network settings.[9] Together with the related discipline of cultural studies, media studies has been a discipline which has invariably confined questions of identity (individuality and 'the subject') as well as questions of power, ideology and community to the great model of language and the frameworks of understanding that have derived from the influence of the 'Copernican revolution' in the humanities inaugurated by the work of the Swiss linguist Ferdinand de Saussure at the turn of the twentieth century (see Chapter 2).

With the exception of a few theorists writing throughout the period of the dominance of media studies such as Marshall McLuhan, Guy Debord and, to a certain extent, Jean Baudrillard, there was very little attention given to questions of form and medium.[10] It was as though the fascination with the content of 'the image' and the discourses surrounding it had somehow concealed the very modes of connection which gave them circulation. Some areas of communication studies, in particular positivist and behaviourist perspectives,[11] have examined the interactive processes which are deemed to exist between two speakers – and dyadic models of communication analysing the relation of sender, receiver and message abound (see Chapter 2). However, the social implications of the actual structures of communication mediums (network and broadcast) have received relatively little attention (save exceptions such as the above).

From the early 1990s onwards, a few years after the Internet began its now infamous exponential growth, the theoretical necessity of analysing the social implications of communication 'mediums' had become paramount, if not unavoidable. It was as though, by the turn of a key, there had been a transformation in the opportunity to understand the integrative dimensions of media that aren't subordinate simply to linguistic derivatives. It was as if media studies had been waiting for an historical object – the Internet – in order to acquire the appropriate lens for understanding communication as medium.[12]

The consequences of this theoretical period of change were that, firstly, some of the early 'medium' theorists like McLuhan and Innis began to be, and are still being, reclaimed (see Chapter 3). Secondly, new distinctions are being made to reflect the renewed importance of distinguishing between 'form and content' such as 'ritual' versus transmission accounts of communication. The understanding of communication as 'ritual' is a radical paradigm shift from the hegemonic status of 'transmission' views of communication, which all but saturated communication theory for the most part of the twentieth century. Put simply, ritual views of communication contend that individuals exchange understandings not out of self-interest nor for the accumulation of information but from a need for communion, commonality and fraternity (see Carey, 1989). Following this approach, transmission models of communication, on the other hand, view communication as an instrumental act – the sending and receiving of messages in ways which individual actors are largely in rational control of.

The latter model of communication, which has in the main dominated communication theory, has been critiqued, either implicitly or explicitly, by philosophers of language who have attacked the identitarian, essentialist, 'logocentric' and 'phonocentric' underpinnings of such a model (see Wittgenstein, Lyotard, Kristeva, Lacan). The project of Jacques Derrida, for example, has been to criticize the idea that language affords a stable stock of meanings for which it is the job of any particular communication to convey. To characterize communication in this way, as *'a transmission charged with making pass, from one subject to another, the identity* of a *signified* object' (Derrida, 1981: 23), is to make all kinds of metaphysical investments in the derivation of meaning and the privileging of communication agents as rational, autonomous selves. These assumptions are radically criticized by Derrida and we will return to them in trying to understand the way in which he claims they are tied to variations in contexts of communication. At the same time it will be possible to see how Derrida's work is also celebratory of a second media age, because the latter's apparent open-endedness unmasks the 'metaphysics of presence' that is able to operate in the more restricted (but never totally) contextual setting of broadcast forms of communication.

However, for the most part, whilst philosophical 'deconstructions' of essentialism are instructive, they have also, it is argued, been overstated. Instead of only examining the way meaning works within texts, this book will focus on how technological infrastructures of communication also need to be examined for an understanding of forms of connection, social integration and community. These material changes, it is argued, also offer a challenge to essentialism, and make it harder to sustain. Hence the need for communication theory which can not only challenge the 'media studies' paradigm, but also show how it is coming to be recast. At the same time, however, media studies, as a theoretical domain concerning itself with the first media age and as harbinger of 'content analysis', remains relevant to the fact that broadcast and the nature of spectacle in modern society are integral to social organization in advanced capitalist societies.

The first and second media age – the historical distinction

The commitment to the idea of a 'second media age' is one that had been gaining ground by the middle of the 1990s with an array of texts – some utopian, others pessimistic concerning the rise of Internet culture and the concomitant demise of broadcast or 'media' culture. Such literature, exemplified by the publication of Mark Poster's book *The Second Media Age* in 1995, has exhibited either a kind of enthralled fascination with the liberating social possibilities of new technology, or, conversely, has encouraged us to rethink what older technologies mean for social processes. But the idea of a second media age had been gaining ground during the 1980s in embryonic form within rubric notions of the information society which was somehow different from simply 'media society'. Indeed the discipline of 'media studies' has become far more ambiguous as its object of study has been made much more indeterminate by the transformations that are currently underway. The term 'media' itself, traditionally centred on the idea of 'mass media', is addressed in the United States by the discipline of 'mass communications'. But media studies (and mass communication studies) in its traditional form can no longer confine itself to broadcast dynamics, and in contemporary university courses it is being subsumed by the more generic scholarship of communication studies – where the accommodation of the distinction between first and second media age is able to be best made.

However, the formalization of the distinction between these two kinds of era has, I would argue, received its greatest momentum in the wake of the domestic take-up of the Internet from the early 1990s. Since that time we have seen a plethora of literature taking over bookshop shelves dealing with everything from technical guides to interactive computing to numerous interpretive texts about the influence the Internet will have on our lives. It is also implicit in a range of journalistic writings in the mid-1990s including Howard Rheingold's *The Virtual Community* (1994), George Gilder's *Life After Television* (1994), Nicholas Negroponte's *Being Digital* (1995) and the corporate musing of Bill Gates in *The Road Ahead* (1996), but also in other, more critical texts like Poster's, Sherry Turkle's *Life on the Screen: Identity in the Age of the Internet* (1995), Pierre Lévy's *Cyberculture* (2001) and various collections like Steven Jones' *Cybersociety* (1995) or David Porter's *Internet Culture* (1997), culminating in the compilation of readers by the late 1990s (Bell and Kennedy, 2000; Gauntlett, 2000; Lievrouw and Livingstone, 2002; Wardrip-Fruin and Montfort, 2003). Not surprisingly, a 'new media age' had also come to feature in numerous texts regarding media policy, in claims that broadcast was rapidly dying and that regulation of digital media forms presented the only remaining policy challenge (see, e.g., Steemers, [1996] 2000). At the same time the heralding of a 'new Athenian age of democracy' by Al Gore, and Third Way political advisers in Britain, became very audible.[13]

By the end of the 1990s the second media age thesis had rapidly become an orthodoxy, and entered the mainstream of New Media thinking.

In Australia, for example, Trevor Barr's account of the Internet, 'Electronic Nomads: Internet as Paradigm' (Barr, 2000), exclaims: 'The Internet's extraordinary growth and global reach of the platform in recent years, the passion of its adherents and its maze of unresolved issues all qualify it as a paradigm shift' (117). Whilst wanting to specify whether or not the Internet will offer 'promise or predicament at the dawning of a new communications era' (144), Barr maintains:

> An inherent strength of the Internet is its anarchy compared to the estab-
> lished modes of ownership and control of traditional media: there are no
> direct equivalents to the 'gatekeepers' of content and form which charac-
> terize the major media of the past few decades, the press and broadcast-
> ing. Everyone who has access to the Net can become their own author,
> expressing their own sense of identity to other Net users scattered through-
> out the world. (143–4)

Even non-specialist media thinkers like Manuel Castells (1996) have taken up a version of a second media age thesis as a critique of McLuhan, arguing that the onset of cable and digital television audiences has brought about more personalized and interactive media culture: 'While the audience received more and more diverse raw material from which to construct each person's image of the universe, the McLuhan Galaxy was a world of one-way communication, not of interaction' (341).

It is the 'interactive society' which has replaced such a world, accord-ing to Castells, in the wake of a symbolically transitional period of 'multi-media' which has given way to a 'new system of communication, based in the digitized, networked integration of multiple communication modes' (374). Castells claims that only within this integrated system do messages gain communicability and socialization: All other messages are reduced to individual imagination or to increasingly marginalized face-to-face sub-cultures. From society's perspective, *electronically-based communication* (*typographic, audiovisual, or computer-mediated*) is *communication*' (374).

Castells is saying that whilst non-electronically based communica-tion may still exist, it is progressively losing its status. This makes access to the 'interactive society' a crucial question, as the world becomes divided into the 'interacting' and the 'interacted':

> ... the price to pay for inclusion in the system is to adapt to its logic, to its
> language, to its points of entry, to its encoding and decoding. This is why it
> is critical for different kinds of social effects that there should be the devel-
> opment of a multinodal, horizontal network of communication, of Internet
> type, instead of a centrally dispatched multimedia system, as in the video-
> on-demand configuration. (374)

These characterizations have not changed much from the arguments of the early to mid-1990s. Early second media age thinkers, Poster, Gilder,

Rheingold, Negroponte and Lévy, are quite coherent in expressing the way in which they claim that the Internet (and interactive technologies in general) enables quite a radical departure from prior forms of social bond. For them the Internet is redemptive in the way it is said to liberate individuals from centralized apparatuses of information, be they state- or corporate-controlled, as exemplified by television. George Gilder (1994), who prides himself with having predicted the demise of television and the birth of the telecomputer as far back as 1989 (101), singles out television, 'the Cathode Ray Tube' and the wireless technology of radio as instrumental in the formation of a pervasive medium empire, the '"master–slave" architecture' of 'a few broadcast centers' that 'originate programs for millions of passive receivers or "dumb terminals"' (26). By contrast 'the much richer, interactive technologies of the computer age' will enhance individualism and creativity rather than mass culture and passivity (23, 32). For Negroponte (1995), decentralization is a major feature of what he calls the post-information age.[14] In providing an alternative to the homogenizing structure of broadcast communication, the Internet is said to offer almost unlimited democratic freedom to track down information, to correspond with thousands of other enfranchised individuals and spontaneously form virtual communities which would not otherwise be possible.

For Lévy (2001), the Internet is a 'Universal without Totality' (91–103), creating a knowledge space where, '[a]s cyberspace grows it becomes more "universal" and the world of information less totalizable' (91). But one of its most important aspects is that it provides an alternative to mass media, to 'communications systems that distribute organized, programmatic information from a central point to a large number of anonymous, passive and isolated receivers' (223).[15]

This model of decentred association is said to be seductive for thousands of consumers who have access to the Internet insofar as it spectacularly overcomes what is seen to be the tyranny of the first media age – broadcast media. Where broadcast media are characterized as a relation of the one to the many, as one-way, centralized communication, they are said to be fragmentary of (geographic) communities in denying interactivity and homogenizing cultural form.

For Poster and Rheingold, who are examined more thoroughly in Chapter 3, an analysis of the architecture of cyberspace relations shows – they claim – that the newer, extended electronic public sphere defies the kinds of instrumental and monopolized centralized control that have traditionally been accompanied by practices of normalization and regulation wrought by broadcast (Rheingold) and the culture industry (Poster). This view persists in much of the second media age literature despite the fact that the Internet has itself become a frontier of monopoly capital.[16]

Compared to broadcast forms of media, the Internet is said to offer free-ranging possibilities of political expression and rights of electronic assembly which encounter far fewer constraints, whether technical, political or social. The celebrated democratizing character of the Internet is

Table 1.1   The historical distinction between the first and second media age

First media age (broadcast)	Second media age (interactivity)
Centred (few speak to many)	Decentred (many speak to many)
One-way communication	Two-way communication
Predisposed to state control	Evades state control
An instrument of regimes of stratification and inequality	Democratizing: facilitates universal citizenship
Participants are fragmented and constituted as a mass	Participants are seen to retain their individuality
Influences consciousness	Influences individual experience of space and time

rooted in its decentralized technical structure. Based on 'packet-switching', a technical network system developed by Rand Corporation in the 1960s, messages, images and sounds on the Internet are always sent in a fragmented fashion by way of multiple routes. This principle was Rand's solution to information held in a database being destroyed in military conflict. Information is always on the move, fluctuating between decipherability and indecipherability and indeterminate in its mobility. Because of this the Internet cannot be controlled either technically (by hackers or programmers) or politically (by states or corporations).[17] In the twentieth century, which was characterized by the control of broadcast apparatuses by governments and corporations, the Internet was also popularly seen to represent an unlimited technical medium for the reconstitution of a 'public sphere'. As Table 1.1 suggests, the public sphere enabled by the second media age restores a two-way reciprocity that is otherwise seen to be denied by one-way communications of broadcast. In addition, the constituency addressed by broadcast is constructed as, and so regarded as, an undifferentiated and largely indeterminate mass, whilst on the Internet the individuality of communicants is redeemed.

In this historical typology, the periodization of an 'age' or era of interactivity – the digital age, the age of the Internet or the second media age – is almost always contrasted with a dark age of mass media.[18] It is a particular expression of an historicist discourse on technology which fetishizes the new and accentuates any differences there might be from the old.[19]

The critique of broadcast is remarkably coherent, whether it be from liberals concerned with public choice and free speech (like Gilder, 1994; Negroponte, 1994; and Rheingold, 1994) or from those employing Marxist frameworks (post-Frankfurt School), or postmodern concerns for the rhizome (as in Deleuze) or the shadow of the silent majority overcoming the simulation machine (Baudrillard, 1982).[20]

Celebrants of the Internet herald its claimed democratic and redemptive virtues either as being able to re-establish lost communities through interactivity or as making possible new kinds of community that transcend modern forms of state control. To quote from Poster (1997), who is working from a broadly postmodernist point of view, the Internet connotes 'a democratization' of subject constitution because 'the acts of discourse are not limited to one-way address and not constrained by the gender and ethnic traces inscribed in face-to-face communications' (222). This is to be contrasted with the broadcast media as a medium of centralized, unilinear communication: 'The magic of the Internet is that it is a technology that puts cultural acts, symbolizations in all forms, in the hands of all partici-pants; it radically decentralizes the positions of speech, publishing, film-making, radio and television broadcasting, in short the apparatuses of cultural production' (222).[21]

Further, insofar as the electronically produced space of the Internet displaces institutional habitats, it breaks down hierarchies of race, gender and ethnicity (see Poster, 2000: 148–70). By allowing the construction of oppositional subjectivities hitherto excluded from the public sphere, the Internet's inherently decentralized form is heralded as its most significant feature – allowing the collision and superimposition of signifiers and semiotic worlds in which the some sense of an authoritative meaning – a *logos* or a grand narrative – can no longer be sustained. This, Poster argues, allows the Internet to subvert rationalized and logocentric forms of political authority, which has imbued the European model of institu-tional life since the Middle Ages. As cyberspace identities are experienced in much more mobile and fluid forms, the public sphere enlarges in the midst of state apparatuses but, at the same time, acts to undermine statist forms of control. This tension is partly played out in those state-originating anxieties concerned as much with the encryption of information against cyber-terrorism as with the use of communications technologies in surveillance.

Broadcast mediums and network mediums – problems with the historical typology

The conviction that we are coming to live in a post-broadcast society, envisaged in the claim that the Internet is going to eclipse broadcast media, is one that has been made by journalists and cyber-theorists alike. The idea that an entire communicational epoch can be tied to key technologies – print technologies, broadcast technologies or computerized interaction – is central to making the distinction between the first and second media age. The distinction is relative rather than absolute, as we shall see, owing to the fact that the significance of the interaction promised by the second media age is defined almost exclusively against the said rigidity and unilinearity of broadcast.

At an empirical level, the distinction between the two epochs is supported by statistics regarding the rapid take-up of interactive CITs, to the point of eclipsing immersion in broadcast environments. There are, however, two problems which come immediately to the fore in tying these epochs so closely to both the innovations in technological development and the take-up of these technologies by consumers of all kinds.

Firstly, all of the various celebrants of the second media age thesis overlook continuities between the first and second media age which, if recognized, would, I argue, shake up many of their social and political claims. However, we should not throw out the distinction between broadcast and interactivity altogether; as we shall see in Chapters 4 and 5, this is an indispensable distinction for a form analysis of modes of communicative integration.

Nevertheless, the second media age thesis does not acknowledge just how much interactive CITs share some of the dynamics of broadcast that they have supposedly transcended, and, what is more, the degree to which they are dependent on and parasitic of broadcast. These continuities, which are addressed in Chapters 4 and 6, involve the way in which CITs, whether we are speaking of interactive or broadcast, operate with similar logics as technologies of urbanization. Secondly, they both produce economic imperatives which are mutually reinforcing, rather than distinct. When looked at from an economic perspective, we shall be able to see how both the Internet and television, network media and broadcast media, 'need' each other.

A second difficulty with the historical distinction made by second media age theorists is the particular alignment of the two epochs with what are seen to be monumental technological developments. It is as though the various possibilities of communication are positively determined by the technology itself (a tendency toward technological determinism) rather than by the recursive relation between technical, political, social and economic environments. Then there is the necessity to distinguish between the structure of communication environments (decentred, centred, one-to-many, many-to-many) and the technical forms in which this structure is realized. Broadcast can be interactive as much as interactivity can be facilitated within broadcast.

Television, print, radio, the Internet and the telephone provide for elements of broadcast and interactivity; it is just that these are realized differently, and at different levels of embodiment in different 'techno-social' relations.

Broadcast can be any form of public spectacle or public address either technologically extended or not; i.e. a lecture amplified by a microphone or not. Interactivity can be technologically extended (the Internet) or simply face-to-face. From the point of view of technologically extended forms themselves, we can also speak of a co-presence of different kinds of media formation. Thus, the significance of the Internet is not that it is a more powerful medium than other channels, but that it provides a platform whose sub-media contain both broadcast and interactivity.[22] Tanjev Schultz has observed that 'on the Web some sites ... become more popular than others. Then they serve as "mass media" on the platform Internet that allows for all kinds of media and types of communication' (2000: 208). Also, those Internet sites which are mirrors of professional established media, such as newspapers, simply add to the original reach which the publication or broadcast already has (see Schultz, 2000: 209).

It is not, therefore, the technologies themselves which bring about these properties in a direct correspondence to a medium. A capacity for broadcast is inherent in a range of technological forms, from the telephone to writing. At the same time the simulation of presence is just as possible in computer-mediated environments as it is with cinema and television.

The fact that so many of these examples have considerable histories to them makes the New Media discourse on 'convergence' theoretically flawed. Convergence is already immanent in old as well as new technologies, but mainly through their interrelation with technologically extended social relations in general.

However, the principal basis upon which convergence is presented as a New Media phenomenon is related to digitization. A review of the history of media and telecommunications technology shows that digitization is not a necessary condition of the convergence of broadcast and network architectures. Convergence may increase the inter-operability needed to access both architectures from one individual portal, but this has much more to do with the historically produced demand for personalization. Nor does digitization particularly privilege interactivity and network over broadcast, as the second media age theorists maintain. Rather, as we will see in Chapter 3, both these theses place technology before any understanding of the anthropology that is at work in contemporary communication environments.

To clear up these confusions caused by what might be called New Media historicism, I argue, in this book, for the need to characterize 'the second media age' not as an epochal shift but as a level of communicative integration which is in fact not new at all but is internal to a range of communicational mediums which have co-existed with broadcast long before the Internet. Brian Winston's instructive history of means of communication from telegraph to Internet illustrates this fact well (Winston, 1998). That new technical mediums somehow have their own aesthetic and social qualities which are separated from 'outdated' mediums is, he reveals, a common misconception resulting from the fetishization of the 'new'.

Winston shows, for example, how economic factors, rather than technology, imposed the primary limitations on the bandwidth of cable communication in the last century. But political and ideological factors which saw broadcasting as a 'centralizing social force' (Winston, 1998: 307) were also instrumental in eschewing cable. Throughout all of the time in which wireless broadcast prevailed, however, 'the wires never really went away', 'the early radio and television networks were wired and the transoceanic telephone cables have kept pace with the development of the international telecommunications satellite system' (305). For Winston, the networks are as old as telecommunications itself, and the inflated claims about the potentials of simply linking computers together are *relatively* hyperbolic.

Nevertheless, for cultural and historical reasons, the arrival of the Internet has 'institutionalized' the idea of network as a normative 'medium', and in doing so it has allowed some theorists to rethink broadcast also as a medium. The term 'second media age' is useful to the degree that it implies a cultural shift in perception toward media environments – insofar as network structures of communication have become much more visibly prominent since the emergence of Internet communication. As we shall see in this book, the turn to reality TV genres away from narrative programming is a part of this shift. Insofar as even broadcast mediums, in a limited sense, also provide a kind of network between communicants – a network based on ritual – the rise of the Internet as a concrete and tangible network allows us to see this.

One of the major reasons why media analysts tie individual technologies so closely to communicational qualities is to do with the way in which CITs are largely empiricized. The significant relationship is seen to be that between the technological doorway to a medium and the consumer. This doorway is one to which we are said to have either an active or passive relationship – typified by the Internet and television, respectively. George Gilder (1993) proposes, 'TV ignores the reality that people are not inherently couch potatoes; given a chance they talk back and interact'. At the 'interface' level of interaction, this might be referenced to the consumer's control of the remote control, which is seen to be relatively passive, as opposed to control over the mouse, which is seen to be active.

In the case of the Internet consumer as opposed to the television consumer, there is an appearance of control over the interaction. This illusion of control is one in which a technology is reduced to that of 'reproduction' (Jones, 1995) – the reproduction of forms of life based on less technologically constituted modes of exchange like the face-to-face and writing. Here, when experienced as a 'use-technology', the Internet is seen to be very much instrumentally subordinated to the carrying on of a social contract by more technically powerful means. The individual who is idealized as participating in this contract is the embodied subject, whose embodiment is somehow overcome and extended. In being a TV consumer, on the other hand, the idea that there is an embodiment to extend is more ambiguous. Instead it is through our selectivity of the channels of messages that we experience that we can participate in pre-constituted modes of life in a technologically extended way.

However, whilst this distinction between activity and passivity can be held up in the situation where CITs are thought of as technologies of reproduction (as tools, or instruments of extension), it weakens considerably when they are accorded the role of technologies of production (as networks and electronic assemblies).[23] The idea of a CIT of production refers to the consideration of information mediums as *environments* (see Meyrowitz, 1999; Poster, 1997) constitutive of altogether new kinds of behaviour and forms of identity.[24] That is to say, they are not just reproducing existing kinds of social relations, but bringing about new ones.

Interaction versus integration

> Media of communication … are vast social metaphors that not only transmit information but determine what is knowledge; that not only orient us to the world but tell us what kind of world exists. (Carey, 1972: 285)

The distinction between activity and passivity as well as that between mediated and unmediated communication falls well within the interactive paradigm, based as it is on the face-to-face or 'transmission' analogue for communication. This long-standing preference in communication theory for the transmission model can largely be attributed to the prevalence of 'interaction' as its basic communicative building block, from which are built the various accounts of communication.

The emerging alternative account is to distinguish between interaction and integration. In this distinction, interaction is still important, but needs also to be viewed in terms of the fact that all concrete interactions occur in the context of dominant frames of communicative integration – which is enacted through abstract 'rituals' of communication (see Chapter 5). The integration thesis rejects the idea that the study of communication is reducible to documenting empirically observable kinds of *interaction*, be these interpersonal or extended.[25] In tribal society, for example, face-to-face relations, and the significance of the body in communication rituals, envelop the social whole. This is observable from the point of view of the rituals and categories of seeing the world that are developed within such forms of social tie (i.e. the anthropomorphizing of animals and objects in the natural world). A person formed within this setting does not actually have to engage in constant face-to-face interactions in order to be enveloped by the set of relations that are bound up in its ontology. Even when such interactions are not occurring, the *ontology* of the face-to-face as a centre of cultural formation comes to frame all other forms of interaction. So, distant forms of communion are made over 'in the image' of face-to-face. Similarly, if we take technologically extended forms of communication as characterizing a social tie of a different order again, we might say that in modern media-saturated societies, mediums like television or the Internet frame our lives even when we are not viewing or using them. This does not mean that we avoid face-to-face relations, or are 'addicted' to technologically mediated interaction; rather it means that we conduct our face-to-face relations 'through' the dominant mediums or social interchange. Here are some examples:

- When we watch a soap opera, we typically are viewing countless thousands of face-to-face interactions between talking heads, whilst, in the very act of such viewership, we forgo our own engagement in face-to-face interaction. Most of the needs we might have for the face-to-face may be achieved via the screen.
- Studies show that people in the city, who have much more access to high volumes of face-to-face contact, use the telephone far more than do people in rural areas.
- Studies of Internet relationships show that anonymous interactants are more likely to divulge intimate information, as if they had a long-term face-to-face relationship, than they would with strangers in embodied interactions.
- Commonplace in the etiquette of Internet communication is the use of 'emoticons' as a substitute for the gestural communication that interactants feel is lost in the medium.

The prominence of the way in which technologically extended communication has become a dominant mode of integration can even mean that we may idealize some kind of unmediated face-to-face sense of community as a reaction to the pervasiveness of extended forms of 'communication at a distance'. Conversely, we might also fetishize communication technology itself as being capable of delivering us the interactive immediacy that is denied to abstract kinds of community (the dream of virtual reality). These two kinds of reactions to contemporary media integration can also be found in much of the more populist variety of second media age literature and cyberstudies texts which privilege the concept of interaction.

Such literature is framed by a social interaction model – i.e. that face-to-face interaction is being supplanted by extended forms of communication – and this is seen to be derived from technology somehow intervening and separating individuals from some 'natural state' of interaction which is the face-to-face. This powerful model inspires not only nostalgic communitarians, such as Rheingold, who claims that individuals in information societies are looking for ways to get back to that which they have lost – the face-to-face – but also postmodernists, like Félix Guattari, who, while sharing the view that face-to-face relations are no longer significant, sees in this no cause for lament. Instead, he argues that it is important to embrace post-individual networks of communication, and realize that the subject is a fiction and always was (see Guattari, 1986). But this kind of negative theology is, I would argue, merely parasitic of the misconception that the face-to-face *was* ever historically lost in the first place. That is to say, if the face-to-face is considered as a form of social integration rather than interaction, these kinds of political oppositions become, I would argue, untenable. It is because, anthropologically, the face-to face is an important mode of connection in information societies that the Internet becomes such a powerful mode of connectedness – but one that can never consummate the mode of integration it supposedly stands for.

Integration and ritual models, on the other hand, look to the kind of background communicative connections which provide the hierarchy of *agoras* of potential assembly, be these public, institutional or virtual, which are independent of individual communicative acts. The crucial point here is this independence. It is necessary to understand how, even when we are not watching television or listening to the radio, the broadcast communication environment still frames our individual lives. We can experience the telephone as though it is an extension of the face-to-face, or, conversely, we engage in the concrete act of face-to-face communication and yet we are somehow 'away' on the telephone or the Internet, only kind of half-present because, really, it is extended forms of communication that are mediating even how we experience the face-to-face. This latter thesis, that the dominant background connections or mediums by which a given group of individuals are socially integrated come to mediate other levels of interaction, is one persistently explored throughout this volume.

In working through this argument, the pertinence of distinguishing between a first and second media age is appraised, and alternative models of understanding how broadcast media and interactive network media are related to each other and to social reproduction will also be presented.

Notes

1   This is why Schwoch and White are concerned with 'an analysis of the pedagogy of technological determinism in American culture' (101).
2   The process of learning the electronic life and the importance of the everyday is a matter to which I will return in the final chapter on telecommunity.
3   This claim is made for both traditional 'images' (see Gitlin, 2002) and New Media (see Postman, 1993; Virilio, 2000). The idea of a 'saturated self' is also central to this (see Gergen, 1991).
4   See the innovative article by Karin Knorr-Cetina, 'The Society with Objects: Social Relations in Postsocial Knowledge Societies' (1997). Knorr-Cetina puts forward an 'end of the social' thesis in referring to the process of 'objectualization' in which increasingly 'objects displace human beings as relationship partners, and embedding environments, or that they increasingly mediate human relationships, making the latter dependent on the former. "Objectualization" is the term I propose to capture this situation' (1).
5   In information societies, the intensity of kinship relations and face-to-face relations has declined in a number of ways. Families are getting smaller and more people live alone. But even the nuclear family, as in the case of Schwoch and White, is increasingly characterized by technological mediation, if not technological constitution.
6   Throughout this book, the term 'the Internet' refers to the 'network of networks' which has been globally standardized since 1991. Although many other CMC systems which facilitate Internet Relay Chat, email, newsgroups, bulletin board systems, MUDs and MOOs may not be, strictly speaking, part of the Internet, as Wellman and Gulia (1999: 189 n. 3) point out, they are rapidly becoming connected to it.
7   Some of the papers produced by the Centre for Contemporary Cultural Studies, such as Stuart Hall's influential essay 'Encoding/Decoding' (Hall, 1980), took as their departure point a critique of the process model. Hall, in a later interview, explains that he first gave the paper at Leicester University, where the communications programme was particularly dominated by process pedagogy (Hall et al., 1994: 253).

8   This debate, between Marxist and postmodern forms of media studies, agreed about the importance of discourse, but conceived entirely different *ends* to their analysis, with Marxism interested in the role of ideology in the reproduction of a social totality, and postmodernism ontologizing the contingency of discourse itself as a denial of totality. Certainly the sociological merits of the Marxist approach would prove to be limited by remaining within the linguistic paradigm. By the 1990s Christopher Williams was asking, is it not 'the case that ideology has become a hopelessly unusable term?' (cited in Corner, 1997: 453). Indeed, in the face of New Media, it can only be wielded as a 'clumsy club', whereas it once had a central role in the unification of media studies.

9   For example, John Hartley (1992a) adheres to a basic theoretical tenet 'that communication is textual, not behavioural' (14). The other tenet he gives is that communication is 'social, not individual'. It is because Hartley conflates all communication with broadcast, or, at least, with understandings that an analysis of broadcast most often yields (the book in which he wrote this is about television), that he overemphasizes texts. My modification of this tenet is that the textual or behavioural qualities of communication are conditioned by the architecture of the medium in which it is realized. I agree with Hartley's second tenet, but the social nature of communication once again has to be related to the means and media of communication. The social is not some abstraction which can be posed over and against the individual, or the means of communication in which individuality is realized.

10   I would suggest that it is because of this imbalance, rather than the incommensurability of different approaches, that media studies has developed what John Corner (1997) has called a 'knowledge problem'.

11   Positivism and behaviourism each subscribe to instrumental views of technology, which are based on a stark separation between the human and the technical. For example, positivist methodologies tend to talk about the 'use' of technologies, 'the user perspective' or rational choice perspective, in which a technology is reduced to a most visible and tangible element, e.g. that on the Internet we use a mouse and make choices. Alternatively, behaviourist models come from the opposite direction in which the individual is rendered entirely passive – their aim being to examine the 'influence' that technology has on individual (only sometimes social) behaviour.

12   But this does not mean that 'media studies was nearly dead', as Gauntlett extravagantly claims in hailing 'long live new media studies' (Gauntlett, 2000: 3); rather, traditional media can be looked at in a new way.

13   For an assessment of Gore's proclamations on the Internet and the 'techno-communitarianism' of Demos, the New Labour think-tank in Britain, see Robins and Webster (1999: 229–31).

14   On decentralization see pp. 157–9. For Negroponte, the post-information age refers to a post-broadcast age of an 'audience the size of one' (164), where information is extremely personalized and not distributed in homogeneous volumes.

15   Moreover, 'Cyberspace … is based not on such a hub-and-spoke model of distribution but on one of shared spaces where everyone can have his say' (223–4).

16   Studies indicate that the same gigantism that afflicts the old media now dominates the new. Despite the Internet's myth of indestructible diversity, cyberspace is also vulnerable to monopolistic tendencies. '[In 1999], 60 percent of all time spent on the internet was on sites owned by 110 companies. By 2001, fourteen companies captured the largest share of the user's time and 50 percent of all time is spent with four companies' (Buzzard, 2003: 207).

17   See Chapter 3. A list of useful guides to the technical details of the Internet is given in Jones (1995: 8).

18   As Silverstone (1999) observes,

> The new ideology of interactivity … [is] … one which stresses our capacity to extend reach and range and to control, through our own choices, what to consume, both when and how. It is hailed to undo a century of one-to-many broadcasting and the progressive infantilization of an increasingly passive audience. It is an expression of a new millennialism. These are the utopian thoughts of the

new age in which power is believed to have been given, at last, to the people: to the people, that is, who have access to, and can control, the mouse and the keyboard. (95)

19 However, some recent correctives to this orthodoxy criticize 'information revolution' as hyperbole, and the modernist myth of the new. Bolter and Grusin (1999) show how processes of 'remediation' of older media by newer media (e.g. TV remediating film or photography remediating painting) are not exclusive to a digital or post-broadcast 'era'. For Winston (1998) the term 'revolution' is wrongly applied to 'New Media', as he proposes to show how the pace of change today is actually slower than in previous periods of technological diffusion and transformation in the means of communication. The 'Information Revolution' is 'largely an illusion, a rhetorical gambit and an expression of technological ignorance' (2).

20 The broad contours of this critique are already anticipated in Bertolt Brecht's short reflection on 'the radio as an apparatus of communication' ([1932] 2003).

21 There is a great deal riding on these claims, stakes which broadcast corporations themselves are now interested in. Geoff Lealand (1999) argues that studies in the USA are being conducted by media corporations, who have commissioned sociologists and communications analysts to study this decentring, and are part of strategies for more comprehensive forms of deregulation.

22 However, this does not mean that the Internet should be seen as producing the same 'field of recognition' as television. For example, some have tried to depict the Internet as television with millions of channels, and millions of broadcasters. The problem is that each channel is weakened in its broadcast power the more channels there are, diluting the exposure of any message or persons who become its 'content'. As we shall see, it is impossible to be famous on the Internet.

23 An overemphasis on CITs as technologies of the production of 'new' social relationships can be seen to be a precursor to the advent of 'complexity theory' – the idea that volume and speed of emergence of causal interconnections between social (or physical) phenomena become so complex and chaotic as to produce new and sometimes chaotic behaviours and properties. (For a postmodern expression of this phenomenon as it applies to communication processes, see Kroker and Weinstein, 1994.)

24 Nowhere is this more spectacular than in the widening generation gap that is emerging between net-literate youth and not-as-literate adults, especially in school classrooms. There is a burgeoning amount of literature in the education journals relating to this (see Downes and Fatouros, 1995; Green and Bigum, 1993; Holmes and Russell, 1999; Russell and Holmes, 1996).

25 Most typical, for example, of the humanist anthropology and behavioural traditions of communication research (see Finnegan, 2002).

TWO

THEORIES OF BROADCAST MEDIA

It is not possible to understand the central dynamics of network communication, or why the second media age thesis has become an orthodoxy, without understanding the nature of broadcast as a *medium*. In fact, as we shall see, the two communicative forms can be argued to be, in the contemporary period, mutually constitutive. That is, I argue, they are mutually related in their practical reality and are also related therefore in how we should understand them.

Understanding broadcast and network as distinct communicative architectures also entails making some fundamental distinctions about the kinds of communication effects which are *internal* to them. The distinction between 'transmission' versus 'ritual' communication is one which provides a useful way of classifying the different kinds of perspectives on broadcast media which emerged in the twentieth century. These perspectives correspond to qualitatively different kinds of communicative processes which are evident in the mass media, and which broadly correspond to content versus form, respectively. The transmission view is by far the predominant one, and is only recently being criticized from the point of view of its overstatement. Instructively, the impetus of this rebuttal is not to be found in the large body of critical writings[1] but can be found in the rise of new kinds of communicational realities which expose transmission views of broadcast as inadequate. The critical literature on 'transmission' views of community has been led in recent decades by a number of French theorists, exemplified by the work of Jacques Derrida, discussed in detail in Chapter 5.

What this and the next chapter aim to do is to introduce the main perspectives on broadcast and network cultures of communication respectively before going on to look at the way in which the perspectives on broadcast need to be critically reassessed. This will mean that shortcomings of instrumental perspectives will become apparent in light of an understanding of network communication, but, in later chapters, we shall also see how broadcast can be seen to carry very important forms of reciprocity and community, *contra* the claims of many of the second media age thinkers.

The media as an extended form of the social – the rise of 'mass media'

The massive changes wrought by the industrial revolutions that have unevenly transformed the developing world have represented important preconditions to the formations of populations living in conditions of density whilst at the same time connected by the framework of the nation-state. The sheer scale of population increases within modern nation-states combined with the migration of people from pastoral regions to cities has created metropolitan densities conducive to the maturing of so-called 'mass society'. Infrastructures necessary to service such growth have led to the mass production of transport and goods, the mass delivery of education and of course the 'mass media' (see Giddens, 1990; Thompson, 1995).

In the period of the breakdown of traditional societies characterized by a high intensity of integration by religion, the fragmentation of nationally framed polities by way of urbanization, the separation of individuals from feudal means of production and the creation of labour-power as a commodity collectively gave rise to a range of perspectives on the 'massification' of society ranging from mass/elite frameworks to liberal-pluralist ones.[2]

The mass/elite framework had its most salient beginnings from the 1930s onwards, which was also the time when the media were first 'mapped out as a field of study in a formal or academic sense' (Bennett, 1982: 38). It was at this time that the co-emergence of cinema and radio combined with rising unemployment and mass armies of disposable workers which culminated in the Great Depression. What all of these frameworks have in common is the idea that the masses once formed by the aforementioned disintegrations are, in late modernity, in need of a mechanism of incorporation for social integration to occur. This may be politically, by way of the gradual enfranchisement of successive groups, or economically, by, for example, the law of value operating in the market to facilitate equivalence between labour-power and commodities. At the same time, however, the mass society framework of the 1930s gave rise to a concern for 'effects analysis' which focused on 'stimulus' and 'response' and the influence that 'the media', deemed to be somehow external to the formation of a person's identity, comes to exert over that identity and culture in general.[3] These studies oscillated between celebrating the media as agents of the education of the masses to condemning them for hypodermically injecting audiences with 'propaganda'.[4] Most of the empirical research was concerned with what people 'think' as a result of being influenced by the media. On some rare occasions, the 'mass psychology' of the media was also studied, such as when, in 1938, H.G. Wells' famous novel *The War of the Worlds* was broadcast in radio form on CBS, resulting in the now difficult to understand apocalyptic hysteria over a Martian invasion.

The mass/elite model of society has been criticized by Marxist perspectives on communication and more recently within cultural studies. The Marxist critique labels mass/elite theory as an ideology of erasing a politics of class (neutralizing the realities of the ruling class versus the working class), whilst cultural studies is concerned with the way in which the framework treats audiences as 'passive'.[5] Interestingly, the Marxist and cultural studies critiques dismiss 'mass society' perspectives insofar as they are deemed to be serious contenders for a sociological framework. Tony Bennett argues, for example, that as a theory of society, it is generally imprecise, that its historical commitments are at best romantic and at worst vague, and that there is no account of the transition between periods of social integration (Bennett, 1982: 37). Yet it is, of course, precisely because it developed in the period when broadcast media were in ascendance that this 'imprecise' theory came about. My own argument is that the mass society outlook, if thought about in relation to the media, is an entirely appropriate response to the embryonic dynamics of media-constituted integration. I agree with the above critiques that it cannot be taken seriously as a sociological framework, but as a theoretical expression of, as well as response to, the way broadcast media are able to reconstitute social relations it provides some early conceptual tools for this – even if these are inadequate by today's standards.

For example, mass society theory is sometimes accused of homogenizing media forms themselves. As John Hartley suggests, 'it is difficult to encompass the diversity of what constitutes print, cinema, radio and television within one definition' (in O'Sullivan et al., 1994: 172). But this is only true if we are interested in the *significatory* properties of these media.[6] Where these media do converge, however, is in the capacity to act as bearers of a homomorphic *medium* of communication, which produces audiences whose field of recognition is vertically constituted.

It is significant that it was only during the period of the massive rise of broadcast through television in the 1950s and 1960s that literature again began to appear dealing with the age of the masses (see Bell, 1962; Kornhauser, 1960; Shils, 1957). This is the time when another, very different kind of mass society theory made its debut in the form of what Stuart Hall has called 'American Dream Sociology'. This kind of sociology, represented by the writings of Daniel Bell, Seymour Lipset and Edward Shils, argued that the general liberalization of society, supposedly measured by the participation of the working class in politics and the growth of welfare, had solved earlier conflicts arising within civil society to the point where a new consensus had been achieved by which resources were at last being distributed according to a harmonious pluralist pragmatism. This thesis, known as the 'end of ideology' thesis, argued that the fundamental political problems of the industrial revolution have been solved: the workers have achieved industrial and political citizenship; the conservatives have accepted the welfare state; and the democratic left has recognized that increase in overall state power carried with it more dangers to freedom than solutions for economic problems (Lipset, 1963: 406). The 1950s renaissance of mass society theory was therefore one with 'the elite' subtracted from it where the masses had been redefined as the melting pot of democratic evolution. Shils was working earlier than the other theorists at revising the 1930s formulations in which the masses had achieved the long march from the outskirts of the social, cultural and political landscape to the democratized and pluralized community or universal speech. Such speech was, of course, guaranteed rather than truncated by the mass media. It is as if in fact such a democratization of the masses had not been possible without the rise of the media. In this way, American Dream Sociology saw the media as simply a transparent extension of the democratic public sphere, a continuation of the social by other means, where the media act in service to the community. As Stuart Hall (1982) describes it, 'in its purest form, pluralism [American Dream Sociology] assured that no structural barriers or limits of class would obstruct this process of cultural absorption: for, as we all 'knew', America was no longer a class society. Nothing prevented the long day's journey of the American masses to the centre' (60).

Contrary to the way in which the presumed homogenizing function of the media was celebrated, several of the empirical studies of a behaviourist and positivist kind conducted at the height of this perspective confirmed the opposite effect, that audiences were in fact highly differentiated and heterogeneous (e.g. Lazarsfeld and Kendall, 1949).[7] Such studies were effectively repositioned by Shils in yet another twist in the tale of mass society theory, as proof of the confirmation of the 'homogeneous' pluralistic tolerance of mediatized democracies.

What is characteristic of both the early and later versions of mass society theory is their adherence to empiricist and positivist epistemologies of the media. That is to say, in arguing that the media are able to extend the democratic process,[8] by circulating views, a number of metaphysical commitments are made which have since been critiqued by linguistic perpectives on the media (semiotic, structuralist and post-structuralist). The media are largely assumed capable of providing a transparent reflection of reality (language is transparent), whether this be as a reflection of events (the news), of culture (popular culture), or of morality and art (film and literature). Secondly, the status of the individual is unproblematic for this model. For example, the position (*qua* perspective) from where a media product might be consumed is disregarded. Thirdly, all individuals (subjects) are deemed to have the same opportunity for observation.

Mass media as a culture industry – from critical theory to cultural studies

A major counter-perspective to the liberal-pluralist idea that the mass media are a democratizing extension of social forms is represented in the

Marxist tradition of the critique of ideology as well as the critique of the unequal ownership and control of the means of communication according to class divisions in capitalist societies. The critique of ideology, which will be explored in the following section, views the media as a powerful apparatus for 'ideologies' – which are not simply just ideas – for reproducing the values and structures that are active in the maintenance of class inequality. But the media are also significantly an industry in themselves, an industry in which commodities are bought and sold.

As the markets and innovations for developing subsistence commodities become exhausted, modern capitalism has tended to turn its attention to industries for which demand has fewer limitations, and has targeted altogether new needs that are created by historical circumstances. Service industries, military industries and leisure industries (tourism, music, entertainment, sport) each provide economic markets which are potentially unlimited and insatiable. The earliest thinkers on this phenomenon were Theodor Adorno and Max Horkheimer, who, in the mid-1940s, published their now canonized critique of the culture industry: 'The Culture Industry: Enlightenment as Mass Deception' (1993).

The culture industry carries all of the hallmarks of capitalist production. Its products are standardized, emptied of aesthetic merit and capable of mass production, and they are consumed on scales as vast as those on which they are produced. The primary consequence of this massification of culture was, for Adorno and Horkheimer that it had profound implications for aesthetic reception. Art is appreciated not for its special ability to communicate truth or beauty but for its marketability. A Hollywood movie has to have a sex scene and a car chase, done in a certain way. The contemporary novel must have a minimum number of elements in order to be a 'best-seller'. The weekly 'life' magazine must have the requisite revelation on weight loss, improving sex life or overcoming relationship and family disorders. But it is not just the conventions *within* genres that become standardized; new genres appear which even mock the masses they are purporting to represent, such as the spectacle of humiliation characterizing 'candid camera', celebrity-hosted talk shows, 'world's funniest home videos' or 'funniest advertisements', or even 'world's dumbest criminals'. Conversely, celebrities have their own television genres, like 'Lifestyles of the Rich and Famous' or 'Entertainment Tonight'. Alternatively, serious social issues like AIDS, Third World relief or the environment receive modest attention, unless they are promoted by a music or film celebrity. From the period when the control of information, communication and entertainment is concentrated in the hands of a few to be sold to the many, culture itself can become a commodity in all kinds of forms.

Insofar as culture becomes massified through broadcast principle, Adorno and Horkheimer see it as replacing religion and the smaller units of integration of the feudalist world. This thesis at its broadest is therefore continuous with the mass society tradition in accounting for the social acceptance and role which broadcast achieves.

To take their opening claim:

The sociological theory that the loss of the support of objectively established religion, the dissolution of the last remnants of precapitalism, together with technological and social differentiation and specialization, have led to cultural chaos is disproved every day; for culture now impresses the same stamp on everything. Films, radio and magazines make up a system which is uniform as a whole and in every part. (Adorno and Horkheimer, 1993: 30)

But the culture industry does not only produce standardized content; it also produces the audience itself by way of 'a circle of manipulation and retroactive need in which the unity of the system [of the production and consumption of meanings] grows ever stronger' (31). This formulation places emphasis on the fact that broadcast produces content for audiences at the same time as it produces audiences for the content – one of the first statements of how the media themselves are a system of social integration which, despite its function as servile to the needs of commodity capitalism, nevertheless facilitates a common culture. In other words the mass is constituted by broadcast; it is not some kind of pre-given amorphous body that has broadcast imposed on it.[9]

For Adorno and Horkheimer, perhaps the most significant feature of the culture industry is that it inculcates 'obedience to hierarchy' (38). In the very structure of the few producing on behalf of the many, it discourages the mass from taking initiative or from questioning the initiative being taken by the elite. It is little wonder that the culture industry produces a loss of individuality (see 41) – a phenomenon which mass society theory, as we saw, does not so much describe as promote in its selection of methodology.

Interestingly, the culture industry thesis shares with the liberal-pluralist perspective the idea of the media as an extension of social relations. However, where there is a fundamental disagreement is over what exactly is extended, which for the Frankfurt School is a replication of obedience to hierarchy continuous with pre-media social relations. Moreover, for them, the mass media collude in the reduction of social life to the flat, one-dimensional intellectual and emotional habits of commodity consumption, thereby completing the process of the spiritless circulation of commodities.

The media as an apparatus of ideology

For contemporary Marxist perspectives on the media, the culture industry is an 'industry' in itself, but is less important as a site of the production of 'new' social relations that might be exclusively derived from mass media than it is as a site of the reproduction of existing social relations – particularly class divisions, but also the divisions of gender, ethnicity and race. The Marxist approach is therefore interested in the meanings that are negotiated within the media, and its influence in the reproduction of forms of consciousness that accord with the reproduction of capitalist social relations. In this section, we will therefore be surveying the idea of 'ideology' as the content of broadcast apparatuses rather than as implicated in the very structure of broadcast, which will be examined in the next section.

Whilst Marxist perpectives largely subscribe to the argument that the media offer an extension, by reflection, of social relations, this is only so in a distorted form. In a class society, it is quite normal that the 'true' character of social relations, of power and of inequality, is misrepresented. In class societies, wealth is distributed away from its producers, but, more importantly, this process is usually masked in some way. This, at least, is the 'false consciousness' argument of orthodox Marxism – the earliest Marxist formulation of the concept of ideology.[10] The 'false consciousness' thesis posits ideology as a distorted, inaccurate representation of the world, which is cultivated by the ruling class and its managerial servants against the interests of the working class. This early formulation persists today in the ongoing concern that some Marxists have with the 'ownership and control' of broadcasting and, in particular, its recent globalized form.

However, this theory has been widely criticized as being based on a correspondence theory of truth – the notion that ideas should transparently reflect the 'real' world. In fact this doctrine of false consciousness has many more continuities and affinities again with liberal-idealist conceptions of ideology than with later Marxist and cultural theory.

In Marx and Engels' writings a number of more sophisticated senses of ideology appear, which were subsequently developed by twentieth-century Marxists for studying media.[11]

Firstly, there is the idea of 'commodity fetishism', a definition found in Marx's later work which laid the ground for a theory of what Georg Lukács was later to call 'reification'. Unlike 'false consciousness', which some Marxists have attempted to apply to all kinds of class society, Marx's theory of fetishism is specific to the capitalist mode of production.

In turning to Marx's major late work *Capital*, we find a conception of ideology that is related to a fundamental distinction between essence and appearance. In *Capital*, economic relationships as experienced in everyday life do not 'reflect' or correspond to the underlying structural mechanisms of which they are an effect. Here, the appearance of capitalism as it actually presents itself obscures from individuals the systemic inner forces which govern their lives. The important point here is that the misrecognition of the 'true' character of social relations is not a 'defect' of the subject; rather it is a result of how social relations *present themselves*.

Thus in Marx's discussion of the fetishism of commodities in Volume I of *Capital* the fact that individuals exchange their labour-power (as a commodity) for other commodities is experienced as an equal exchange around which an entire realm of legitimation is erected – what Marx calls the 'noisy sphere, where everything takes place on the surface in full view of everyone' (Marx, 1976: 279; see also Hall, 1977: 324). Marx argues that the 'essence' of commodity exchange is really an (abstract) exchange of labour, the source of social value, whilst to the individuals who exchange this labour this only ever appears to them as the concrete relations between things (in the form of price). Whilst this obscures the social character of labour, this essential reality is displaced to the sphere of exchange, which becomes all the more real: 'To the producers, therefore, the social relations between their private labours appear as *what they are*, i.e. they do not appear as direct social relations between persons in their work, but rather as direct social relations between things' (Marx, 1976: 166–7, my italics). From this it may be seen that the 'appearance' is in a sense 'real', especially because it is convincing. Real as it may be, Marx reminds us that it *conceals* the essence, an essence which explains the appearance and an essence which is not manifest to individuals: '... by equating their different products to each other in exchange as values, they equate their different kinds of labour as human labour. They do this without being aware of it' (166–7). In other words, it is not necessarily *ideas* which represent the world 'inaccurately'; rather it is the nature of capitalism itself to present itself in an inverted form.

In terms of the distinction between content and form that is to be examined in relation to the media, Marx's account of the commodity is instructive. Later we shall see how it has influenced the work of Jean Baudrillard and Guy Debord, in which the media themselves, in the form of signs, become intrinsically bound up with the exchange circuits of commodities. In fact, for Baudrillard and Debord, the world of image and spectacle becomes the ultimate form of commodity reification. This important concept, which had its first comprehensive development in the work of Lukács, denotes a phenomenon in which the relations between individuals are said to acquire a 'phantom objectivity, taking on autonomous, all-embracing and rational relations between things (Lukács, 1971: 83). The production of commodities comes to dominate the whole of society constituting appearances consisting of complexes of isolated facts. It permeates the division of labour within the state, bureaucracy, industry and especially science.

The final sense of ideology in Marx and Engels to be examined here is that of ideological incorporation, formulated in their book *The German Ideology*:

> The ideas of the ruling class are in every epoch the ruling ideas, that is, the class which is the ruling *material* force in society is at the same time its ruling *intellectual* force. The class which has the means of material production at its disposal, has control at the same time over the means of mental production, so that thereby, generally speaking, the ideas of those who lack the means of production, are subject to them. (Marx and Engels, 1970: 65)

The understanding of ideology that is purveyed here is one in which the ideas of one group, the ruling group, become generalized to the whole of society. This is often interpreted as a mechanical relationship. But as this was later developed, the fact that one class may monopolize the means of mental and material production does not guarantee that it can simply *impose* its ideas; rather, these ideas are negotiated in a way in which their rule is accepted.

This stance on ideology was developed by the Italian Marxist Antonio Gramsci – by way of the concept of hegemony. This refers to an ideological struggle in which the ruling class compromises with the working class in return for its leadership in society as a whole. It is a consensual form of power in which Gramsci identified the mass media as central. This does not require direct editorial control of media by the capitalist class; rather, managers, who identify politically and ideologically with the ruling class, provide 'the organic intellectuals' who are at the front line of hegemonic struggle.

In the Gramscian framework of hegemony, 'false consciousness' is a myth in that people are seen as having '"true" conceptions in their heads of society as it actually presents itself' (see Alford, 1983: 8) – that is, they have 'common-sense' experience of exchange relations and the division of labour. Therefore 'direct' human experience is the point of origin, the source of their 'real' conceptions, which explains why they acquiesce in their conditions, as 'there is no conceivable alternative to the commodity-form' (Alford, 1983: 7). Thus, individuals' 'common-sense' experience of the world tells them not only what exists but also what is *possible*. In this framework, ideology is merely a more systematic version of common sense, which legitimizes doctrines of particular social groups involved in the organization of the presentation of hegemony. Gramsci problematizes the doctrine that ideology is only ever an expression of class interests (and so an individual's ideological position can be 'more or less read off' from their economic position) as being far more contradictory, and he sees class relationships as potentially more fragile.

For Gramsci, the dominant classes don't merely prescribe ideology for working-class consumption; rather, they have to continually strive to limit the boundaries of the making of meaning to exclude definitions of social reality which conflict with *their* horizon of thought – the struggle for hegemony is won and lost not just in the media, but in the institutions of civil society (such as the family, the churches, the education system, but also in more coercive apparatuses: the law, the police, the army, etc.).[12]

Gramsci's examination of the institutions of civil society was taken up in the 1960s and 1970s by the French Marxist Louis Althusser, who reworked the analysis in developing a very strong link between ideology and the power of the state. Althusser claimed that ideology, and what he called the 'ideological state apparatus', had become much more important in the twentieth century than the repressive and coercive state apparatuses of the nineteenth century. This change could be attributed to the important addition which Althusser makes to the state apparatus, which is the apparatus of broadcast.

Interestingly, it is not merely apparatuses of communication that are important here, but also those of the structure of ideological processes which occur in all institutional settings of power – religious, educational, political and workplace. For Althusser, the growth of electronic broadcasting institutions (particularly visual broadcasting) merely consolidates the consensual integration of individuals that occurs in the *structure* rather then the content of ideology.

In what follows, we shall investigate Althusser's radical departure in thinking on the nature of ideology from both the early Marxist and liberal notions. His innovation involves questioning the very notion of what it means to be an individual in a communication process, an innovation which has been echoed ever since in the analyses of what today is called 'post-structuralism'.

Ideology as a structure of broadcast – Althusser

Althusser's most striking point of departure from the humanist Marxists is to question the categories in which ideology is thought. Ideology is not found in the content of messages, nor is it adopted in the consciousness of individuals; rather, it is nothing less than the mechanism by which the individual experiences selfhood – as an autonomous knowing subject in a world of knowing subjects.

Althusser's account of ideology is almost a reversal of the conventional humanist accounts. For Althusser, there is no such thing as 'given' individuals with an experience of the real; rather, the very idea of individuality is created in the communication process itself. By his account, this process by which the individual is constituted only intensifies in the age of 'mass media'. Indeed it makes possible the 'cult of the individual' which Émile Durkheim first discussed at the turn of the twentieth century.

For Althusser, individuals (subjects) are never essential but are constituted (an 'effect' of ideology). The centrepiece of his theory is his distinction between the individual and the subject. His major proposition in this regard is that '*the category of the subject is only constitutive of all ideology in so far as all ideology has the function (which defines it) of 'constituting' concrete individuals as subjects*' (Althusser, 1971: 160). In other words, Althusser is not denying the existence of individual 'personality'; it is just that such 'personality' is only possible in and through a communication process. The mechanism by which this occurs he describes as 'interpellation', where he says that 'all ideology hails or interpellates concrete individuals as concrete subjects' (162).

For Althusser, ideology only exists by the subject and for the subject, and its function is to constitute people as subjects. While it may seem 'obvious' that individuals are unified, autonomous beings whose consciousness and unique personality are the source of their ideas and beliefs,

Althusser maintains that this obviousness only comes from people '(mis)recognizing' themselves in the way that ideology 'interpellates' them, calls them by their names and in turn 'recognizes their autonomy' (162). It is in this imaginary misrecognition that the subject is *constituted*; the subject is therefore formed in an imaginary relation – 'it cannot be the pure subject of the empiricist notion of experience because it is *formed* through a definite structure of recognition' (Hirst, 1976: 387). Ideology does not constitute individuals in a singular divine act; rather, 'ideology has always-already interpellated individuals as subjects'. For Althusser, *individuals are always-already subjects* in the same way that ideology itself is 'always-already' known (Althusser, 1971: 175–6).

As 'autonomous' subjects with a unique 'subject-position' in the social formation, individuals willingly 'work by themselves' (181) as a 'centre of initiatives' (182). However, whilst the subject is a 'centre of initiatives' responsible for its actions, it is also a subjected being who submits freely to the authority of the Subject – God, Father, institution, the boss, etc. – that is, *a subject through the Subject and subjected to the Subject.*

> The structure of all ideology, interpellating individuals as subjects in the name of a Unique and Absolute Subject, is *specular*, i.e. a mirror-structure, and *doubly* specular: this mirror duplication is constitutive of all ideology and ensures its functioning. Which means that all ideology is *centred*, that the Absolute Subject occupies the unique place of the Centre, and interpellates around it the infinity of individuals into subjects in a double-mirror connection such that it *subjects* to the Subject. (Althusser, 1971: 168)

*   *   *

Althusser's theory represents something of a paradigm earthquake for the study of broadcast media and its social significance. In suggesting that, firstly, ideology is not simply a moment of signification but is the very condition by which it is possible to act as a self-conscious subject, and, secondly, that structures of interpellation which exhibit specular and centred structures are the most significant sites of ideology, broadcast media become an extremely important kind of state apparatus. Althusser's theory points to a sense in which ideology – what he calls ideology-in-general – can be considered a structure of broadcast rather than just content. Ideology as content he refers to as ideology-in-particular. For Althusser, particular ideologies may change but ideology-in-general is an enduring structure. This is why, as Sprinker (1987: 279–80) has argued, the behaviour of media audiences should be seen not as psychological but as social.

Because, for Althusser, ideology is the very condition of a subject being a subject at all, he argues that no one in any society can do without ideology – without a representation of themselves as subjects, of their world and of their relation to the world. This is why ideology is not merely a representation of people's conditions of existence (distorted or otherwise) but rather is 'a "representation" of the imaginary relationship of individuals to their real conditions of existence' (162). For Althusser, as a Marxist, the political point of this statement is that in a social formation where production relations (and inequality) are obscured, where conditions which govern people's existence aren't manifest to them, 'they necessarily live these absent conditions in an imaginary presence "as if" they were given' (Hirst, 1976: 386). Therefore ideology is active in maintaining the status quo of the existing relations of production – active in the reproduction of social relations. However, as we shall see in Chapter 5, Althusser's theory is also important for an understanding of forms of social integration which can be seen to be quite independent of the needs of the reproduction of capitalism.

The society of the spectacle – Debord, Boorstin and Foucault

The power attributed to ideology-in-general in social integration and social reproduction provides a useful theoretical backdrop to understanding the 'spectacle' thesis in French media theory – in particular the theories of Guy Debord and later Jean Baudrillard. This thesis also argues for the basic externalization and objectification of social reality in the media, but it is less a function of narrative than it is of the role of spectacle in the generation of a world of simulation. Their theory is a post-representational one in which the fact of the image rather than what the image says becomes the most important aspect of present-day broadcast societies. The system of images transforms the mundane into a hyperreal carnival of totemic monuments through which the 'masses' achieve congregation.

Debord, Boorstin and Foucault

In understanding the significance that is attributed to the image in the various theories of spectacle, it is important to specify the fact that 'the image' derives its power almost exclusively from the medium of broadcast. We will see in the next chapter that, with the Internet, there is no such thing as 'the image', as the Internet does not provide a field of visibility in the same way as broadcast does. The image is a function of media in which there is a concentration of the attention of the many on a particular monumental event or representation. When such representations are repeated over time – when images become icons – the image is able to take on a life of its own – where the things it refers to become secondary. Indeed the referent may disappear altogether.

    An early and original theorization of the phenomenon of the reification (cf. Lukács, above) of the image in modern society is given in Guy Debord's well-known monograph *The Society of the Spectacle* (1977). First published in France in 1967, this text takes a 'situationist' perspective on broadcast media. Debord's argument is that capitalist culture presents itself as an immense assemblage of spectacle. But spectacle for him is not just 'a collection of images, but a social relation among people, mediated by images' (epigram no. 4). The spectacle even promotes itself as an agent of the unification of society as a whole. It is the domain of society which 'concentrates all gazing and all consciousness' (epigram no. 3).

For Debord, the modern media, for which he contends the term 'mass media' is a 'superficial manifestation' (aphorism 24), are agents both of political power and of urbanization. They secure the complacency of the population to inequality and hierarchy:

> The oldest specialization, the specialization of power, is at the root of the spectacle. The spectacle is thus a specialized activity which speaks for all the others. It is the diplomatic representation of hierarchic society to itself, where all other expression is banned. (aphorism 23)

At the same time the spectacle is a practical agent for the dual unification and separation of individuals around the principle of private consumption:

> The spectacle originates in the loss of the unity of the world, and the gigantic expansion of the modern spectacle expresses the totality of this loss: the abstraction of all specific labor and the general abstraction of the entirety of production are perfectly rendered in the spectacle, whose *mode of being concrete* is precisely abstraction. (aphorism 29)

Debord describes the situation of the spectacle – as simply one representation of the real – splitting off and separating from the real as though it has transcended it:

> The spectacle is nothing more than the common language of this separation. What binds the spectators together is no more than an irreversible relation at the very centre which maintains their isolation. The spectacle re-unites the separate, but re-unites it *as separate* (aphorism 29).

In Debord's account, a view which is restated by Fredric Jameson (1991) nearly two decades later, the image is – following a somewhat Lukácsian trajectory – presented as 'the final form of commodity reification'.

Six years prior to Debord's publication, across the Atlantic, the phenomenon was receiving theoretical attention in the form of Daniel Boorstin's publication of *The Image* (1962).[13] Boorstin saw television and cinema as an extension of the de-naturing and de-realization of modern society wrought by the electronic management of the environment. In modern society,

> distinctions of social classes, of times and seasons, have been blurred as never before. With steam heat we are too hot in winter; with air conditioning

we are too cool in summer. Fluorescent lights make indoors brighter than out, night lighter than day. The distinctions between here and there dissolve. With movies and television, today can become yesterday; and we can be everywhere, while we are still here. In fact it is easier to be there (say on the floor of the national political convention) when we are here (at home or in our hotel room before our television screen) than when we are there. (231–2)

For Boorstin, broadcast technologies are servile to what he described as the 'homogenization of experience', in which differences between individuals are flattened rather than expressed – leaving individualism itself as the remainder. Nowhere is this more salient than in public opinion polls:

... the rising interest in public opinions and public opinion polls illustrates ... the rise of images and their domination over our thinking about ourselves. ... Public opinion, once the public's expression, becomes more and more an image into which the public fits its expression. Public opinion becomes filled with what is already there. (239, 240)

A consequence of Boorstin's claims is that, in the age of the image, public opinion is no longer able to be surveyed or polled. Polling itself, a positivist gesture of research, cannot quite cope with the fact that it is attending to a thoroughly anti-positivist reality.

Debord and Boorstin's depiction of the social function of spectacle bespeaks striking continuities with that of Michel Foucault's account of public displays and torture in eighteenth-century Europe.[14] J.B. Thompson (1995) gives a good account of this in sketching the formation of modern forms of power:

The societies of the ancient world and of the ancien régime were societies of spectacle: the exercise of power was linked to the public manifestation of the strength and superiority of the sovereign. It was a regime of power in which a few were made visible to the many, and in which the visibility of the few was used as a means of exercising power over the many – in the way, for instance, that a public execution in the market square became a spectacle in which a sovereign power took its revenge, reaffirming the glory of the king through the destruction of the rebellious subject. (132)

Thompson argues that Foucault's work is instructive for a theory of the media, less in his promotion of discourse analysis than in showing how the older spectacular forms of power became manifested in institutional life in routine fashion, imbuing surveillance and disciplinary regimes in an involutory way. That is, the 'disciplinary society' which Foucault details in *Discipline and Punish* is one in which 'the visibility of the few by the many has been replaced by the visibility of the many by the few' (Thompson, 1995: 133).[15]

Of course Adorno and Horkheimer would argue that these two forms of recognition relation are intertwined. That is to say, the visibility of the few by the many which is organized by the few is also the means by which the few are able to control the many through economic and cultural subordination. In an article on television, Adorno (1954) comments: 'the more inarticulate and diffuse the audience of mass media seems to be, the more mass media tend to achieve their "integration"' (220). The gaze of the audience is sold to advertisers, at the same time as *selection* of the content of media programmes is itself highly coded within dominant ideological interests.[16]

In the context of modern mass media, the institutionalization of this commodification of the gaze is one which imposes an entire order of symbolic inequality, in which the masses associate via the image and the celebrity (for further discussion, see Chapter 6).

This inequity in the production of 'cultural capital' that is central to broadcast as a system of reproduction of late capitalist societies occasionally surfaces at the level of discourse. The central operation of the performative nature of broadcast is not itself visible. We know from Althusser that, in fact, the very operation of 'interpellation' and of the calling function of ideology is one that is upside-down. Althusser puts it in psychoanalytic language – that it is conscious on the condition that it is unconscious – but the effect is the same. For Althusser, the structures of the system of interpellation are, by definition, impossible to examine.

However, it is possible to argue that at the level of discourse, the structure of interpellation sometimes surfaces in narratives which, when the analysis of what constitutes broadcast is taken into account, can be seen to be self-referential: an abstract reflection of the medium itself but explainable in terms of the medium. Here are three such discourses.

The discourse of 'ordinary people'

It is only in the media cultures dominated by 'spectacle' that it is possible to speak of ordinary people. The now familiar way in which individuals who do not work for the culture industry, or are not subject to any significant media attention, behave when interviewed by a television network or press or radio is instructive here. A very common narrative of a person being called on to describe their role in an event, a process or in society at large is one which runs, 'I am just an ordinary person doing my job.' But even news narratives replicate this 'interpellation' of the individual in describing how 'ordinary men and women are to be affected by this or that government decision'. Ordinariness cannot simply be explained as some deeply constituted residue of feudal class dynamics in which one's position is more or less determined by birth. The discourses of 'ordinariness' can be seen to closely coincide with the rise of 'mediated publicness'.

It is only under the conditions of the polarization between celebrity and ordinary culture that a film such as *Forrest Gump* could be made. An interesting film from the point of view of defying any easy genre classification, it centres on the character of Gump, who, with humble means and simplistic technique, is able to achieve an extraordinary range of things, from marathon running, to heroic war service, to gridiron stardom. Whilst the film is predominantly concerned with celebrating the culture of opportunity said to underwrite the moral superiority of the United States, it is also about how even the most ordinary person can, in a society of celebrities and spectacle, be noticed and satisfied.

The discourse of 'the system'

In this discourse anonymity is rejected in favour of a reflexive critique of abstract domination. Just as ordinariness has replaced the specification of 'underclass' or working class, something distinctively co-emergent with the mass media, so too the specification of politicians and the ruling classes has been replaced in populist discourse by a rebelliousness to something called 'the system'. A phrase which was taken up by the counter-cultures since the 1960s, it has entered into popular discourses in ways which denote everything from the suffocation of expression and creativity, to the inevitability of domination, to a generalized cynicism of power.

The discourse of 'they'

'They' are building a new freeway. 'They' have discovered a cure for cancer. 'They' are opening a new shopping centre. 'They' aren't telling the public the full story. Perhaps the most pervasive term to accompany the rise of the mass media is that of 'they'. Who, exactly, are 'they'? The fact that the mass which is constituted by broadcast media is indeterminate as far as particular messages go implies that the individuals who are part of this mass are also indeterminate to each other. In other words, broadcast makes possible scales of association which are difficult to achieve by any other means. On the one hand, we can talk about a high level of integration via the image and the celebrity, but, on the other, we see relatively weak kinds of connection at the horizontal level of the division of labour. In media societies, 'otherness' is completely concentrated in the fetish of the spectacle or the celebrity, whilst at the level of the everyday, it is radically diluted. But what kind of other are 'they'?

There are many theses. 'They' is simply a shorthand for the institutional nature of the entity being described – the roadbuilders, scientists and doctors, developers, the government, etc. 'They' could also be a default way of saying 'I can't elaborate on the detail' or 'It is more complex than my description warrants.' 'They' could also simply be an absent-mindedness, a carelessness about 'who' it is that makes the daily news. 'They' might be a polite way of saying also that we can't know 'who' 'everyone' is, nor would we want to. But when considered in relation to the structure of broadcast media, it is clear that, for example, celebrities are not they; their own identity is well defined, so much so that the media produce genres of programming and magazines which are exclusively *about* celebrities. Obversely, they aren't written about and yet seem to be everywhere. 'They' substitutes for the modern loss of specificity. We are not quite sure how it works, we are not invited to participate, but *they* know. 'They', in this reading, is the emblem of individual disconnection and disembodiment – of the fact of the loss of various practical knowledges which are based in cultures of mutual presence and oral culture.

With all of these discourses, the question arises as to whether 'they' are peculiar to broadcast integration or to technologically extended culture in general – of which the Internet is a part. This will be reassessed in Chapter 4.

Mass media as the dominant form of access to social reality – Baudrillard

In the last section we saw how, whilst spectacle has become a highly visible social reality, its influence over social behaviour is not so visible. This influence is nevertheless manifest in specific discourses, which provide rare cases in which the field of recognition created by the broadcast medium condenses into the content of that medium.

The way in which the attention of the audiences is concentrated though spectacle is not unlike a contemporary form of 'reification' of social relationships where the fetish of representations overtakes the conditions of that representation. The spectrality of the image, and the successive forms by which it becomes detached from social relations in general, is also a central concern of media sociologist Jean Baudrillard. But unlike the spectacle thinkers, Baudrillard argues that the ascent of a culture of images produces a crisis in representation itself. In media societies, processes of signification are no longer underwritten by a metaphysics of presence or the promise of recovering some kind of original, authentic or privileged meaning.

The eclipse of ontology by the image rests with what Baudrillard sees as the power of 'simulacra'. This term refers to the way in which what we consume from media becomes more real than what it supposedly refers to. In elaborating the evolution of simulacra in his essay 'The Precession of Simulacra' in *Simulations* (1982), Baudrillard takes us through four phases of the representation the image. The image in its different guises:

- is the reflection of a basic reality;
- masks and perverts a basic reality;
- masks the absence of a basic reality;
- bears no relation to reality whatsoever: it is its own pure simulacrum.

Baudrillard examines the cultural status of representation in European society to suggest that it evolves through the above forms of phenomenality.

The first phase is easily recognizable in the code of journalists today, who, with their narrow conventions and frameworks of objectivity, bias and neutrality, embrace the prospect of a correspondence between reality and the representations they produce. The second phase is also recognized in various understandings of ideology discussed above, that representation is largely a distortion of real conditions. The third phase is probably the most difficult to understand. Here, Baudrillard argues that an objective representation of the real is impossible, because the referent is already a simulational reality. Therefore, representation hides not 'the truth' but the fact that there is no 'truth'. His most famous example is probably his claim that the function of theme parks like Disneyland is to encourage us to think that the rest of society is somehow 'real' – whereas for Baudrillard the entire world has today become, in a sense, a giant theme park (see, in particular, Baudrillard, 1988).

The fourth phase marks the end of social reality itself as an available referent. This is easy to understand. The connection to the referent can become lost altogether – something which is indicated by the emergence of a number of interesting genres like 'reality TV'. What is represented on TV is supposed to be more significant than other forms of experience. At the same time, the television itself can be found colonizing our public lives everywhere we turn, in taverns, shopping malls, delis, laundromats, airports, train stations, hardwares and local stores. As McCarthy (2001) argues, 'TV integrates into our everyday environments so well that we barely notice its presence' (2). Indeed, according to Baudrillard, these two senses of the screen becoming the real (the screen colonizes the real, and the real is only 'real' if it is on a screen) mean that images begin to refer to *each other* rather than to the 'real' world.

This relationship is not unlike the kind of relationships involved in commodity fetishism, which Marx investigates. As discussed earlier, for Marx, it is only via the commodity that individuals experience their connection to each other. We can recall that whilst commodity fetishism conceals the 'essence' of the commodity (which for Marx is labour), the 'appearance' of the commodity in the advertisement and on the shelf is also 'real' and therefore convincing.

For Baudrillard, it is the image itself that becomes the measure of all things, including our access to social reality. 'Everywhere socialization is measured according to exposure through media messages. Those who are under-exposed to the media are virtually asocial or desocialized' (Baudrillard, 1983: 96). The image is highly convincing and we do not seem to be able to live without it. But the greater our exposure to the mass of images, the more 'information' we receive, the more we come to live in a world of less and less meaning: 'Information devours its own contents; it devours communication and the social'; it 'impodes', and for two reasons: firstly: '[i]nstead of causing communication, *it exhausts itself in the act* of staging the communication; instead of producing meaning, it exhausts itself in the staging of meaning'. Here Baudrillard argues that meaning is devoured more rapidly than it can be reinjected insofar as information and the image become self-referential – 'a closed circuit' (99). Secondly, the media do not bring about socialization, but the implosion of social relationships in the only remaining relationship created by the mass media – the masses. Insofar as all relationships must 'pass though' the media relationship, they suffer the entropic force that is the condition of simulacra.[17] The implosion of meaning right down to the microscopic level of an individual sign, what a word can mean, is mirrored by this macroscopic implosion of the social, in a way that echoes McLuhan's formula – the medium is the massage.

The 'mass'-age, in Baudrillard's terminology, is an exclusive effect rather than a precondition of the media. The mass and the media are the shadow of each other, and when dynamics of simulacra prevail, that institution known as the 'social' becomes outmoded, absorbed into the image. In this world the individual becomes 'a pure screen, a switching centre for all the networks of influence' (1983: 133), in a world wherein 'we *form a mass*, living most of the time in panic or haphazardly, above and beyond meaning' (1983: 11).

Here the media no longer function as a massive lie in pretending to represent fiction as the real or the real as fiction. What Baudrillard means by hyperreality or simulation is different. There can no longer be a contrast with the real; rather, the real is produced out of itself as the performativity of the mass media is amplified above all other events.

For Baudrillard, the masses aren't the kind of duped underclass that are to be manipulated by the media and politicians (a notable departure from the mass/elite and Marxist frameworks); rather, they are a kind of ground of absorption and massive gravitation which neutralizes all meaning and creates the conditions for a society of nihilism and cynicism.

The masses are a stronger medium than all the media: 'it is the former who envelop and absorb the latter – or at least there is no priority of one over the other. The mass and the media are one single process. Mass(age) is the message' (Baudrillard, 1983: 44).

The medium is the message – McLuhan, Innis and Meyrowitz

The final important perspective on broadcast media that I want to explore is that of Marshall McLuhan and Harold Innis, which Joshua Meyrowitz has called 'medium' theory. Whilst not having as much currency as the 'spectacle' and ideology frameworks, it has recently received a large amount of attention (see Adilkno, 1998; Bolter and Grusin, 1999; Goodheart, 2000; Jordan, 1999; Meyrowitz, 1995, 1999; Skinner, 2000; Wark, 2000). Most of this attention is directed towards seeing McLuhan as a rediscovered prophet of a second media age, but much of it is also interested in an affirmation that it is, after all, important to look at communication media.

McLuhan's work is based on an historical understanding of successive waves of communication from print to electronic. His various aphorisms on the media, including 'the global village' and 'the medium is the message', have become absorbed into popular culture, whilst not necessarily understood within McLuhan's own system of thought. Influential in the academy in the 1960s, McLuhan underwent a 'loss of vogue' (McQuail, 1983: 90) in the 1970s, which continued until the recent reclamation of his work by theorists of the second media age and cyberculture.[18]

The major contribution of McLuhan to communication theory is his multi-dimensional account of communication 'mediums' – a way of looking at technologically constituted social relationships, which each have their distinct reality or ontology. This approach is very different from, say, the culture industry thesis, the theory of ideology, or Baudrillard's precession of simulacra, each of which implies a basic homogenization of those immersed in media.[19]

Rather, McLuhan's contention is that media technologies carry distinct temporal and spatial specificities to which correspond definite frameworks of perception. As James Carey (1972) suggests,

> The exploitation of a particular communications technology fixes particular sensory relations in members of society. By fixing such a relation, it determines a society's world view; that is, it stipulates a characteristic way of organizing experience. It thus determines the forms of knowledge, the structure of perception, and the sensory equipment attuned to absorb reality. (284–5)

Historically, however, he does argue, one or more of these frameworks may come to dominate cultural perception as a whole. Thus, he distinguishes between print-based culture and electronically extended culture. In print culture, claims McLuhan in *Understanding Media* (originally published in 1964), our perception of the world tends to be englobed by literature and the book, which becomes an analogue conditioning other experiences. This is often experienced as the new mediating the old and interiorizing it:[20] ... 'the "content" of any medium is always another medium. The content of writing is speech, just as the written word is the content of print, and print is the content of the telegraph' (McLuhan, 1994: 16).[21]

THE TELEGRAPH
PRINT
WRITING
SPEECH

Today, McLuhan's schema as applied to the Internet might look like the following:

THE INTERNET

PRINT
WRITTEN WORD
SPEECH

IMAGE
ICON
VISUAL COMMUNICATION

These layers of technological worlds, past and present, intensify the work of processing meaning which confronts consumers immersed in the different mediums. This process work becomes heightened to the point where we have to be educated and inducted into it as increasingly information has to be produced by the audience or the receiver.

McLuhan's primary distinction that is relevant here is that between 'hot' and 'cool' mediums.[22] Hot mediums like radio and cinema circulate a large amount of information, bombarding the viewer or listener. Relatively little is required in order to interpret them. Cool mediums, on the other hand, presuppose interaction. McLuhan's assumption is that in hot mediums there is an overdose of information, and there is little need for interactivity, for 'active' participants, or for participation at all.

Later in *Understanding Media* McLuhan begins to describe the demise of mechanical media like print in making way for technologies of 'automation' like radio and television as part of what he calls the 'cybernation' transformation of modern society. It is the electronic instantaneity of radio and TV which consolidates the hegemony of mass media over older mechanical technologies of reproduction.

> Automation brings in real 'mass production', not in terms of size, but of an instant inclusive embrace. Such is also the character of 'mass media'. They are an indication, not of the size of their audiences, but of the fact that everybody becomes involved in them at the same time. (McLuhan, 1994: 372)

In other words, the significant property of broadcast which McLuhan zeros in on is its 'live' character. Here it is the fact that a broadcast communication is *live for the audience,* rather than live at the point of production. The content of the transmission could have been prepared earlier or at the same time as the audience is consuming it. However, McLuhan is, of course, not interested in the content, but in the way the audience is merely a constituted reflex of the medium itself. Insofar as the media achieve cybernation, 'the consumer' of a message also 'becomes producer in the automation circuit, quite as much as the reader of the mosaic telegraph press makes his own news, or just *is* his own news' (McLuhan, 1994: 372). The value of McLuhan's analysis here is that he suggests that an electronic assembly or 'virtual' assembly does not have to be dialogical or equal, or even have 'high participation', in order to guarantee mutual presence. Even if the vast majority of 'participants' in a medium are passive (as in a hot medium), they are nevertheless able to experience mutual presence as the really real.

Most controversial among McLuhan's theories is his later emphasis on the human–technical extension argument where the definition of what qualifies as media is dramatically extended. In a shift from 'the medium is the message' to 'the medium is the massage' (see McLuhan and Fiore, 1967), McLuhan views anything that can extend the body's senses and biological capabilities (psychic or physical) as earning the status of media. 'The wheel is an extension of the foot, the book is an extension of the eye, clothing an extension of skin, electric circuitry an extension of the central nervous system' (McLuhan and Fiore, 1967: 31–41). Whilst, as we shall see in later chapters, there are enormous problems in referring CITs (communication and information technologies) exclusively back to the body in a kind of corporeal essentialism, McLuhan paradoxically allows us to understand recent developments in the *convergence* of CITs with transportational and architectural technologies in a way that is most useful.[23]

The cryptic eccentricity of McLuhan's work overshadowed some of his contemporaries, who, in a number of ways, were more comprehensive and rigorous in their analysis of technical mediums of communication and forms of political power.

One such writer, Harold Innis, presented a medium theory which is perhaps more user-friendly for a theory of broadcast. In *The Bias of Communications* (1964, originally published 1951)[24] Innis makes a major distinction between two kinds of 'empires' of communication. The first, corresponding to the printing press and electronic communication, results in spatial domination (of nations and of populations) – what he calls a 'space bias' – whilst the second, 'time bias', based on oral culture and the cloistered world of the manuscript, accommodates memory and continuity. For Innis, the oral tradition needs to be reclaimed. Broadcast belongs to the empire of space, and in the time he was writing, the early 1950s, it had come to structure prevailing power relations.

As David Crowley and David Mitchell (1995) depict him:

> Innis … saw a recurrent dialectic in History where one medium asserted primacy in a society, followed by efforts to bypass the social power that gathered around the control of that medium … each new mode of communication was associated with tearing individuals and their entire forms of life out of their traditional moorings in locality and place and relocating them within larger and more dispersed forms of influence. With modernity, this process of co-location of the self within multiple spaces, identities, and influences intensifies; human agency itself is progressively pulled away from the local and reconstituted within the expanding possibilities of the modern. (8)

Despite a lapse in the momentum of medium theory in the 1970s, it certainly had some sophisticated exponents in the 1980s and 1990s, among whom Joshua Meyrowitz, whose work is explored further in the following chapters, is exemplary.

Meyrowitz's major work *No Sense of Place* (1985) was a carefully theorized volume which attempted to continue the traditions begun by McLuhan and Innis. For Meyrowitz, electronic media reterritorialize 'sense of place' and the spatial, political and social conditions of this sense of place. They do this by their cross-contextuality and reach, the way in which they can asymmetrically bring together extremely diverse groups who are otherwise separated in cultural focus, in space, and perhaps also in time. Media, especially electronic media, make possible arbitrary relations between a concrete space and a sense of place. By undermining 'the traditional association between physical setting and social situation' the constraints of embodiment such as being in one place at the one time disappear (7). The value of this analysis is in anticipating what has recently only been attributed to 'cyberspace', the mobility that is afforded to an Internet consumer, highlighting the 'virtual' aspects of broadcast.

The value of the 'mediationists', as David Crowley and David Mitchell describe them, is that they were the first to draw attention to the interrelation between different media and systems of power. Their work neither is based on a philosophy of consciousness nor is it behaviourist. In the next two chapters I discuss it further, firstly, in Chapter 3, in terms of how network media have heightened the importance of medium theory; and secondly, in Chapter 4, in terms of how medium theory allows us to theorize the relationship between network and broadcast media.

Notes

1   The major corpus of critical accounts of the transmission view has come from post-Saussurian philosophies of language.
2   Bennett (1982) gives a useful survey of the different perspectives.
3   Effects analysis quickly established itself as a serious pursuit of sociological research. In American sociology, dominated as it was by positivist methodologies, the opportunity for empirical testing of the various theories about media effects presented itself (see McQuire, 1995).
4   For the former function, see, for example, Leavis (1930); for the latter function, see Chakhotin (1939).
5   Oddly, this latter critique is confused about the different kinds of 'mass media'. For example, in a dictionary definition on the subject, John Hartley distinguishes between print, screen, audio and 'broadcast' media. Broadcast is therefore equated with whatever might be in some sense 'live' throughout a signal radiation apparatus (in O'Sullivan et al., 1994: 172–3). Here Hartley is caught up in a cosmology of media 'effects', the study of how the media affect audiences. For example, even in critiquing the idea that the media influence the mass, and arguing that audiences are much more active and intelligent than mass society theory would have us believe, the very prospect of resistance presupposes an effects model.
6   The Marxist and cultural studies frameworks are primarily interested in the way media are industrially and state regulated.
7   More recently audience studies has become a branch of media studies in its own right, which stresses the idea of the active and diversified audience. See Ang (1991) and Gitlin (1998) – the argument that there is no such thing as a single Habermasian public sphere.

8   The question of democracy in the first and second media age is discussed in Chapters 3 and 4.

9   For a 1990s text which empirically investigates the way in which modern audiences are the product of the management and marketing efforts of media organizations, see Ettema and Whitney (1994).

10 Ideology is thus identified with passages from Marx's earlier work in *The German Ideology* as a false, imaginary, upside-down, illusory representation of reality: 'in all ideology men and their circumstances appear upside down as in a *camera obscura*' (Marx and Engels, 1970: 47).

11 The literature on Marx and Engels' concepts of ideology is vast and I will offer here a summary of them to the extent that they are a useful background for examining the media as a state apparatus. For a useful overview of Marxist theories of ideology, see Larrain (1983) and Eagleton (1991).

12 The struggle for hegemony usually entails the manufacture of consensus by way of the revival or construction of deviance. These may be internal to a particular nation-state, such as criminality, counter-cultures and subcultures. External deviance might be projected as a threat posed by other cultures and other nations as either economically, militarily or culturally dangerous – culminating in the modern discourse of 'terrorism'.

13 Given that this text is almost entirely grounded in an empiricist epistemology (see my discussion above), Boorstin manages to capture the import of spectacle in a persuasive way.

14 It is perhaps remarkable that Foucault did not write anything about the modern media, even though he was writing throughout the heights of the society of spectacle. Certainly his work is taken up by media studies and cultural studies in substantial ways, but particularly in the discourse analysis perspective, as we have seen earlier in this chapter.

15 The visibility of the few by the many was not always a matter of violent display, but indeed a common form was also the 'royal progression' which continues today in nations with a monarchical head of state. The regency would conduct routine and regularized regional tours to be visible to his or her subjects on a repetitive basis, as, for example, the British monarch does today, in relation to the Commonwealth.

16 The view that the function of broadcast institutions is primarily to sell audiences to advertisers was first put most strongly by the Canadian Marxist Dallas Smythe (1981).

17 'It is useless to wonder if it is the loss of communication which causes this escalation in the simulacra, or it is the simulacra which is there first.' There is no first term, Baudrillard argues: '… it is a circular process – that of simulation, that of the hyperreal: a hyperreality of communication and of meaning, more real than the real. Hence the real is abolished' (1983: 99).

18 Especially by the editors of *Wired* magazine.

19 For an excellent comparison of Baudrillard and McLuhan, see Smart (1992: 115–40) and Huyssen (1995).

20 Recently this process which McLuhan describes has been taken up in the concept of 'remediation'. See particularly Bolter and Grusin (1999).

21 However, McLuhan often gets these relationships between forms and content confused. For example, in one passage in trying to explain how we positivize the content and ignore the medium, he says: 'The "content" of writing or print is speech, but the reader is almost entirely unaware either of print or of speech' (1964: 26). To be consistant, McLuhan must surely mean, 'the reader is almost entirely unaware either of print or of *writing*'.

22 See the next chapter for a critique of this distinction.

23 For example, McLuhan's outlook is, even if in a limited sense, able to make some of the basic connections which are being made today between 'spaces of flow' (Castells), be these of bodies or messages. See, for example, Meyrowitz (1985), Morse (1998), Graham and Marvin (1996), Calhoun (1992).

24 McLuhan declares his indebtedness to Innis on a number of occasions. In *The Gutenberg Galaxy* (1967) he pronounces: 'Innis was the first person to hit upon the *process* of change as implicit in the *forms* of media technology' (50).

THREE

THEORIES OF CYBERSOCIETY

Cyberspace

Throughout October 1999, concerts were held in London, New York and Geneva to launch 'NetAid', the Internet equivalent to the 'Live Aid' movement of the mid-1980s. The 'Live Aid' movement was comprised of a series of globally broadcast rolling concerts sponsored by corporations who received a moral injection to their advertising profile, as well as patrons at the gates who felt that they were doing something for needy people they had seen on TV.

The later version of empathy-at-a-distance is one in which, by sitting at Internet terminals, those people living in economically and informationally rich countries can do 'something to help'. The Secretary-General of the United Nations was on hand at the concert, to explain: 'Most people in needy countries have to get by on less that two US dollars per day; now, with the click of a mouse, everyone can help. There are no more excuses, let's bring on a new day.'[1]

The heralding of the Internet as universalist and redemptive has, at the turn of the millennium, become a widespread discourse, in which the rhetoric of salvation through an electronic assembly has attained theological proportions. Whether by rhetoric or by clever marketing, the rate of growth of connection to the Internet network is astonishing.

Cyberspace and virtual reality

As suggested in the Introduction, the distinction between the first and second media age is a relative one, and is founded on a heightened contrast between the new network mediums and the structures of broadcast mediums. In this chapter we will explore this contrast by examining 'second media age' thinkers who contend that the growth of the Internet is a reaction to the restricted and unequal possibilities of broadcast. As we will see, there is a surprising degree of agreement from thinkers liberal, Marxist and postmodernist over the emancipatory qualities of the Internet. But before presenting this analysis, it is necessary to explain some of the technical and structural characteristics of new interactive media and assess the claims made for a second media age.

Whilst the term 'cyberspace', which first appeared in the prophetic fiction writing of William Gibson, is most frequently used today inter-changeably with the Internet, some thinkers have pointed out that it can be applied in a much wider sense to include a range of technically consti-tuted environments in which individuals experience a location not reducible to physical space (see Escobar, 1994; Ostwald, 1997).

By this definition, any medium which encloses human communica-tion in an electronically generated space could be a form of cyberspace. A further distinction is also often made to designate that such a space may be very private or shared by others. For example, a personal music listen-ing device with headphones, which Sony Corporation first made famous with the 'Walkman', qualifies as a medium of the enclosure of experience.[2] However, it falls short of the conditions necessary for cyberspace in that it disallows a shared appreciation of the one media 'event'. The event is personalized because its 'performance' and the environment within which it is consumed are connected by an individual user.[3] Thus, the dis-tinction being drawn here can be recognized in a range of daily media habits. Meyrowitz (1985) notes: 'There is a big difference between listen-ing to a cassette tape while driving in a car and listening to a radio station, in that the cassette tape cuts you off from the outside world, while the radio ties you into it' (90).

However, the difference between accessing shared media events and ones that are personally programmed tends to be overlooked by virtual reality theorists insofar as they are preoccupied with bandwidth as a lead-ing marker of its definition. In general, virtual realities tend to require much broader quanta of bandwidth in order to achieve their simulational properties. Thus, virtual reality is regarded as having found a technologi-cal home in digital environments. However, just as peronalization is not an exclusive feature of digital media or a 'second media age', neither is wide bandwidth.

Across the broadcast medium, significant differences exist between the virtual qualities of media. Consider the difference between television and cinema. Cinema offers almost double the bandwidth of TV. An average size television fills 5% of the visual field, whilst the other 95% is occupied by possible distractions in the room. Cinema engages 10% of the visual field, with the other 90% blacked out – eliminating distraction. Cinerama spans 25% of the visual field, whilst virtual screens fill 100% of the visual field as such screens receive their data from computer-generated images. But the technology of projection is merely an extension of broadcast technologies.[4]

As I argue in the Introduction to *Virtual Politics* (Holmes, 1997), unlike virtual reality, cyberspace does not rely on a deception of the senses to create the illusion of an integral realism. Rather, it is by the con-struction of computer-mediated worlds in which (predominantly text-based) communication can occur that an objectivated reality is established which does not depend on a common deception of sense-impressions. As Ostwald argues, 'the critical component of any definition of cyberspace is the element of community', because he maintains that a single person does not exist in cyberspace, but in virtual reality (Ostwald, 1997: 132).

According to James Carey (1995), and, later, Jon Stratton, the most primitive but original place to find the 'origins' of cyberspace is in 'nineteenth-century attempts to speed up circulation time' (Stratton, 1997: 254). Therefore the most fruitful place to look, says Stratton, is to the advent of the telegraph in the first half of the nineteenth century. In the observation of James Carey: 'The simplest and most important point about the telegraph is that it marked the decisive separation of "transportation" and "communication"' (cited in Stratton, 1997: 254). Stratton contends that it is not the emergence of the computer and the microchip per se which inaugurates the production of cyberspace, 'but the increase in the speed of communication over distance to a point where the time taken for a message to traverse the distance reduces to a period experienced by the receiver, and sender, as negligible' (254). By Stratton's reading, therefore, the development of global telecommunication and of cyberspace is inextricably intertwined.

Among the major precursors of computer-mediated cyberspace technologies, the telephone can also be counted. As a twentieth-century innovation on the telegraph, the telephone exhibits virtual kinds of features as an electrically sustained low-bandwidth medium, whilst enabling a limited kind of electronic assembly. Such an assembly, whilst generally only mutual for a few persons at a time, nevertheless facilitates a sense of a meeting place, a place that is augmented by voice mail and answering machine services. The telephone also exhibits a limited number of features of virtual reality insofar as it is semi-enclosed (a given conversion cannot be heard simultaneously by anyone other than the interlocutors) and it translates the voice into a 'meta-signal', electrical pulses which convey analogue sounds. With regard to this latter quality, one of the first theorizations of 'virtual reality' can be found in an early classic on telecommunication by Herbert and Proctor. The second edition of their work *Telephony* (1932) distinguishes electrical current and electrical voltage from what they name as a separate 'virtual' current and 'virtuvoltage'. This distinction is an – albeit crude – attempt to signify the fact that a telephone exchange, in which individuals are jacked in to each other by way of operators or agents, purveys an environment that transcends the purely electrical. This *other environment* stands somewhere between the human voice and the electrical medium, but lacks the comprehensiveness of mediums which today earn the appellation of cyberspace.

Cyberspace and the Internet

The fact that cyberspace is so often conflated with the Internet belies the fact that there have long been other networks before the Internet which qualify as domains of the 'matrix' or cyberspace. The sum total of these networks is sometimes called Barlovian cyberspace, so named after John Perry Barlow (Grateful Dead band member), who applied Gibson's term to CMC as a more complex kind of a space than that which is engaged in a telephone conversation.

Today the Internet has consolidated into a 'network of networks'. Mostly originating from the USA, the major networks which have added themselves to the Internet include ARPANET (government-funded), Fidonet (alternative cooperative), Usenet, the WELL, the thousands of corporate and government intranets, and the World Wide Web. CMC systems that predated many of these networks, such as email, news groups and bulletin board systems, are now carried with the expanded Internet network.

One also needs to distinguish between commercial and domestic networks of CMC. Commercial networks have long predated the domestic, with IBM having its own global intranet some twenty years before the Internet properly began.[5]

Certainly, in America, ARPANET was one of the most instrumental in pioneering the domestic conditions for today's Internet. Built by a Boston company under contract, 150 sites had been established across the USA by the late 1980s. It was designed from the start to allow remote log-in by passwords, a feature that co-developed with the accelerating speed of computer modems in the home.

Of surprise to many of the architects of ARPANET was the fact that one of the most popular sub-media to spring up was email. As Tim Jordan (1999) explains:

> The key point about email is that rather than people using ARPANET to communicate with computers, as the designers expected, people used it to communicate with other people. This was despite the fact that email was not programmed into the system but was added unofficially in an *ad hoc* way. Email emerged spontaneously as the basic resource provided by ARPANET and this has been true of virtually all computer networks. People connect to people using computers, which has given rise to the over-arching term computer-mediated communication. (38)

However, CMC does not just have to be point-to-point, as what the various networks have allowed that was unachievable with pre-CMC communication is correspondence from the many to the many – multiple authors and readers for which there is no technical limitation. Such a form of communication achieves an efficiency impossible in embodied form. Three hundred people can more easily speak to each other in a listserve conference where each message is recorded in a linear sequence of when it was sent (an automatic queue for speech) than could the same three hundred trying to have themselves heard at an embodied conference.

A CMC conference is just one example, therefore, why we should be dissuaded from seeing cyberspace as merely an extension of social relations which occur outside of it, as clearly it is generative of new relations that were not previously possible.

The Internet and its sub-media

However, whilst 'cyberspace' brings about new possibilities of association, the form they take is conditioned by the various sub-media that are available by way of the Internet.

Too often, 'virtual communities' are simply tied to some generic power attributed to 'the Internet'. It is important to specify the various sub-media of the Internet and their implications. As is pointed out by a number of analysts, early fascination with MUDs and MOOs has declined substantially in proportion to the dominant uses of the Net. 'While chat rooms, news groups, and multi-purpose Internet conferences were meaningful for early Internet users, their quantitative and qualitative importance has dwindled with the spread of the Internet' (Castells, 2001: 118).

For Castells, the Internet is not an amorphous ocean that individuals dive into, but a galaxy of regulated sub-media: 'The Internet has been appropriated by social practice, in all its diversity, although this appropriation does have specific effects on social practice itself' (118). Drawing on empirical research, Castells concludes that the on-line identity-building forums available on the Internet are mostly concentrated among teenagers: 'It is teenagers who are in the process of discovering their identity, of experimenting with it, of finding out who they really are or would like to be …' (118).

Castells' observation that virtual communities have an adolescent bell curve contradicts the speculative forecasts of the early 1990s that the Internet can facilitate the formation of very large-scale, so-called 'virtual communities'. These assume the form of voluntary spontaneity without control by a state apparatus as a result of the Internet's web-like structure, a structure which is the legacy of a decentred system of sending information.[6] The mere fact that it is decentred was argued to be the basis for the Internet's alluring emancipation.

The attractions of Internet communication

Of course, the ideological claim that the Internet sets information and its users 'free' was a powerful one in the early years, and was seen by many writers to be the foundation of a new frontier. The frontier image became the reigning metaphor of what David Silver has called 'popular cyberculture', which refers to that period of civic education of populations into the attractions of the Internet (see Silver, 2000: 20–1).

But the horizontal/acentric shape of Internet communication offers attractions that exceed other network architectures (namely, the telephone) – such as bandwidth, the capacity to convey complexity.

This capacity enables also the possibility of sophisticated reciprocity in a way which displaces modes of reciprocity in face-to-face, institutionally extended (where a third person becomes an agent of reciprocity) and electronically extended relations. In making possible more abstract modes of interchange than these other modes, digital reciprocity engenders the paradoxical quality of returning to the historically more unmediated of these modes – the face-to-face as its ideal model – whilst materially annulling this mode as a cultural ground (see the discussion of 're-tribalization' in the work of McLuhan below).[7] The distinctive features of optical fibre, which underpins this capacity, are advertised in its potential for computer, voice, graphics and video services, a more extensive host of media which can guarantee more 'convincing' high-fidelity realism to the user. Such complexity had never been available to analogue forms of electrical transmission, in a way which could be connected up in instantaneous, high-speed and multi-data networking. The instantaneousness of the reciprocity alone is one specific feature which makes possible the metaphorical reconstruction of intersubjective realism – hence the tendency to conflate 'cyberspace' with 'virtual' culture.

The production of what are essentially broadband kinds of interactive environments is qualitatively different from the networks of interchange based on the electric current alone. This is so because the time-worlds and space-worlds – the electronically reified environments – that optical fibre enables are more than merely metaphorical extensions of intersubjective relations but have the potential to replace and redefine the complexity of communication systems. Digitally platformed network communication cannot, like 'the media' (remediated or otherwise) that we explored in the previous chapter, be conceived as a continuation of a system of speech by other means or even a pretence of the same, in the sense that it enables constitutively new kinds of interaction that are arguably historically unique. In particular, the digital nature of this communication places it beyond the function of *extension* which analogue technologies are able to serve (see a longer discussion of this below).

Electrical-analogue time-worlds have never been adequate for the construction of intersubjective simulation systems. It is only by *appropriating the quality of the speed of light, combined with the capacity to convey complexity*, that so-called 'real-time' and near-instantaneous reciprocity are made possible in extended form.[8]

These kinds of technical capacities are also, it is said, remaking the form and content of technologies traditionally associated with broadcast, like television. For example, Sherry Turkle (1995) argues that in the 'age of the Internet', television genres have become much more hyperactive in ways which resemble the random travelling which occurs in cyberspace: 'quick cuts, rapid transitions, changing camera angles, all heighten stimulation through editing' (238), a hyperactive style epitomized by MTV – television's answer to multi-media.[9] This change in tolerance towards a level of freneticness that has become acceptable to television viewers, and now commonplace in nearly every rapid-cycle television advertisement we watch, is mirrored by the fragmentation of the culture industry itself. As Tim Jordan (1999) points out:

> During the 1980s in the USA, the number of independent TV stations grew from sixty-two to 330, while the share of prime-time audience held by the three major networks dropped from 90 per cent to 65 per cent. ... From hand-held video cameras that allow the production of home entertainment to the creation of hundreds of different TV channels, the mass audience that once constituted the consumers of immaterial commodities has been shredded. (158)

To the extent therefore that even traditionally well-defined broadcast technologies are, by convergence with interactive technologies or by diversification, becoming more personalized, more amenable to a sense of active and interactive control by audiences as well as remarkably expanded programming choice, it is argued by second media age writers that a second media age is able to absorb the first media age and reshape it.

However, as we shall see, what such an argument has to contend with is the difficulty of distinguishing between broadcast and interactivity as a purely technical distinction, rather than a distinction resting on forms of social integration.

Theories

The second media age thesis – the Internet as emancipation from broadcast media

As already argued in Chapter 1, the second media age thesis has become an orthodoxy in New Media theory, an orthodoxy which has been taken up almost by default, in many cases with little theoretical engagement or formulation of positions. In what follows I shall focus on the most cogent exponents of the thesis as a way of comparatively appraising its significance in relation to other perspectives.

In accordance with the above observations, the Internet stands out as a comprehensive technoscience world which exemplifies 'cyberspace'. With its large range of sub-media (MUDs, ICQs, email and WWW) and its ability to facilitate complexity, it offers a network medium unparalleled in its potential and scope.

The contention that the Internet and interactive technologies in general have embedded themselves so substantially in the daily existence of individuals living in information societies as to have all but usurped the power of broadcast media is one that is most forcefully put by second media age theorists.

> In film, radio and television, a small number of producers sent information to a large number of consumers. With the incipient introduction of the information 'superhighway' and the integration of satellite technology with television, computers and telephone, an alternative to the broadcast model, with its severe technical contraints, will very likely enable a system of multiple producers/distributors/consumers, an entirely new configuration of communication relations in which the boundaries between those terms collapse. A second age of mass media is on the horizon. (Poster, 1995: 3)

As discussed in the Introduction, unlike theories of broadcast, which have been around for some time, theories of cybersociety or the second media age are, for the most part, very new. Because the Internet, as the most spectacular technology of electronic network communication, has only really globally existed in domestically available form since 1991, communication studies remains in a process of formalizing this new domain of research. The array of theories, from journalistic to academic, has been burgeoning. Like the Internet Revolution itself, the rate of growth in literature about new communication technologies has been dramatic. And as with the pure acceleration of technological change, the literature is characterized by an urgent impulsiveness which produces many generalizations and knowledge claims which become redundant at about the same rate as information technologies themselves.

As noted in the Introduction, since 1991, we have seen a massive growth in computer-related literature. Prognoses of the paperless society and the end of the book have not materialized. Instead, book sales have, if anything, increased, with the weight on each shelf now redistributed to a flourishing computer section.

Apart from the very short history of cyberspace analysis, there is also a much larger body of theory relevant to the second media age from pre-Internet days – theories whose time, it could be argued, has arrived. Of the broadcast media thinkers, the most prominent to bridge the first and second media age are probably Marshall McLuhan, Harold Innis and Joshua Meyrowitz, discussed in the previous chapter. Because content is of far less importance in studying cyberspace, it is not surprising that the medium theorists are able to come to the fore.

On the linguistic side there is the work of Derrida, who, in my view, is the only thinker from the semiotic tradition, apart from Baudrillard, whose work lends itself to a medium theory. The import of the thought of these writers will be dealt with later in the present chapter. But first it is necessary to examine in more detail the claims of the second media age thinkers.

Theorists of the second media age argue that both broadcast and interactive communication apparatuses have together constituted the primary forms of cultural mediation in information societies since the Second World War. The important point here is that it is not possible, in this view, to understand the second media age without understanding the first media age. Traditional media are, as we shall see, central to the distinction between the first and second media age. Writers such as George Gilder in *Life After Television* (1994), Sherry Turkle in *Life on the Screen* (1995) or Mark Poster in *The Second Media Age* (1995) understand the way in which the second media age has arisen on the back of the conditions produced by the first. These conditions – the production of an indeterminate mass by broadcast, the separation of individuals from the means of producing their own contributions to public communication and the disintegration of traditional community – are all hailed as being overcome by the Internet. But it is an exaggeration to suppose this overcoming is a permanent condition or that decentralized network communication simply annuls the power of *centralized communication* apparatuses. Rather, the power of the former is continuously and relatively parasitic on the power of the latter.

According to the second media age perspective, the tyranny that is attributed to broadcast lies in its hegemonic role in the determination of culture (the culture industry) as well as individual consciousnesses (the theory of hegemony) which derives from its predominantly vertical structure. This structure is one in which the individual is forced to look to the image and electronic means of communication to acquire a sense of assembly and common culture. The second media age, on the other hand, bypasses this 'institutional' kind of communication and facilitates – for the romantic variety of cyber-utopians, it 'restores' – instantaneous, less-mediated and two-way forms of communication.

At the level of interaction, the second media age utopians point to the *empirical* increase in the take-up of the Internet and other network technologies, and to the fact that empirically it *is* true that the Internet is mainly interaction and very little broadcast whilst television is mainly broadcast with very little interaction, as evidence for the 'ontological' nature of the second media age as a distinctive trend, movement and modality of social integration. The importance of the fact that the many can interact with the many in cyberspace is almost exclusively related to the way it is said to break the 'lock-out' predicament which individuals face in broadcast interaction. The contraining walls on mediated activities that are erected by the power of broadcast rapidly disintegrate as a form of electronic communication is made available which is adequate in speed, form and complexity to encompass the abstractness of the social forms involved.

In Figure 3.1, the individual is subject to one-way communication from the 'elite' producers of messages. The horizontal connection with other consumers of the same messages is generally only possible via the fetish of the image or the celebrity, in whom (as Durkheim once argued in relation to religion) concrete consciousnesses are concentrated.[10] Conversely, with the Internet, the message producers are bypassed, as the walls that are erected at the horizontal level effectively disappear.

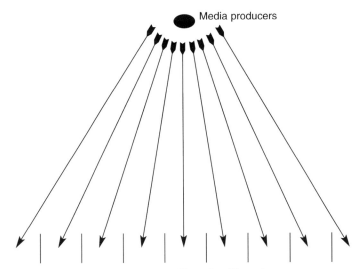

Media producers

Media consumers as indeterminate undifferentiated 'mass'

Figure 3.1   Transmission model: high integration/low reciprocity

These 'media' walls are the result of the architecture of broadcast itself. As we saw with Debord in the previous chapter, the more the individual looks to the media so as to acquire a cultural identity, the less he or she looks 'sideways' for interaction. Conversely, the less the individual looks sideways for social solidarity and reciprocity, the more this mode of association becomes weak and de-normalized, and so the alternative dependence on a centralized apparatus of cultural production becomes imperative.

In the second media age, however, the walls separating individuals at a horizontal level are overcome, as the individual looks directly to others for a sense of milieu and association. As Poster (1995) explains:

> Subject constitution in the second media age occurs through the mechanism of interactivity. ... interactivity has become, by dint of the advertising campaigns of telecommunication corporations, desirable as an end in itself, so that its usage can float and be applied in countless contexts having little to do with telecommunications. Yet the phenomena of communicating at a distance through one's computer, of sending and receiving digitally encoded messages, of being 'interactive', has been the most popular application of the Internet. Far more than making purchases or obtaining information electronically, communicating by computer claims the intense interest of countless thousands. (33)

The Internet lifts individuals out of the isolation created by media walls – particularly as these walls are reinforced in urban contexts. In information societies, individuals increasingly interact with computer screens, developing face-to-screen relations rather than face-to-face relations, but this opposition is no longer significant, argues Sherry Turkle, when the larger cultural contexts of post-industrial societies are eroding the boundaries between the real and the virtual. It is not possible to think of the individual as alone with his or her computer, as Sherry Turkle explored in her 1984 text *The Second Self*; rather, as she more recently suggests: 'This is no longer the case. A rapidly expanding system of networks, collectively known as the Internet, links millions of people in new spaces that are changing the way we think, the nature of our sexuality, the form of our communities, our very identities' (Turkle, 1995: 9). What Turkle describes as the 'Age of the Internet' is synonymous with the opportunity to build virtual communities 'in which we participate with people from all over the world, people with whom we converse daily, people with whom we may have fairly intimate relationships but whom we may never physically meet' (10).

The extent to which the Internet is hailed as an overcoming of fragmentation and individualism is quite remarkable in recent literature. In some cases it is attributed with an integrative function which is able to correct a tendency that is over two hundred years old.

As Dave Healy argues, 'the networked citizen ... is never alone'. To the degree that the Internet represents a 'culture of coherence', he argues, it serves as 'a corrective to the dangers of individualism' which Alexis de Tocqueville spoke of in his visit to America in the 1830s (Healy, 1997: 60).

The message of redemption which is promoted in the second media age thesis, be this for public or private, is a resounding one, a message whose dreams of unity have theological undertones, to which I shall return in Chapter 6. But for the most part, the second media age thesis is derivative of a neo-liberalist broader faith in the emancipatory potential of new means of communication, regardless of the actual exchanges that are encouraged by such means. As Armand Mattelart (2000) has suggested, an 'ideology of limitless communication – but without social actors' has taken over from an 'ideology of limitless progress' (120).

The computer-mediated communication (CMC) perspective

There is an alternative account of electronically extended interactivity that significantly predates the second media age thesis, namely the computer-mediated communication (CMC) perspective.

The CMC perspective overlaps with the second media age perspective but is distinctively concerned with the way in which computer communication extends and mediates face-to-face models of communication. In this perspective the computer is as much a tool as a window onto cyberspace. What it is that gets mediated in this perspective is face-to-face *interaction*, whether this be between two people or many as in a chat group. A symptom of this is the fact that CMC literature is often concerned with how individuals try to develop ways of substituting the absence of face-to-face relations on the Internet: for example, by observing neti-quette[11] (the idea that cyberspace demands forms of polite protocol one would expect in embodied life), or by the growth of emoticons – the symbols used in email denoting facial expressions.[12]

There are four major ways in which CMC literature differs from the second media age thesis. Firstly, it is focused on the uniqueness of the communication event in cyberspace. Secondly, it is concerned much more with *interaction* than with *integration*, that is, the myriad of individual interactions rather than the overall social contexts and rituals by which these interactions become meaningful. Thirdly, unlike 'media studies', some CMC frameworks are interested in how 'external factors' influence a communication event. With broadcast analysis, very little exploration occurs of how outer contexts influence media content; rather, media content is assessed according to how it might reflect or express non-media realities. Finally, whilst not concerned with the kinds of *social* integration which might underpin CMC, it is concerned with *information integration*, the way in which communicating by way of computers is based in information processes that can be found in a burgeoning number of interactions mediated by computer. This latter point opens out the domains of cybernetics and the information society, fields of analysis which can be broadly collected together under the umbrella of information theory.

Information theory The CMC perspective is a continuation of conduit models of communication first discussed in the 1950s. So before looking at the contemporary features of CMC it is worth sketching the main contours of information theory. Oddly, these theories were less relevant to broadcast than they are to dyadic reciprocity – be it face-to-face or electronically extended. The fact that they achieved some considerable influence in the United States during the height of broadcast defies the fact that they were never able to accommodate the phenomena of performativity, of spectacle, and reification examined in the previous chapter. Dyadic models of communication are not very helpful in explaining what happens when a few centres of cultural production send messages to an indeterminate mass.

The main elements of this outlook, some of which have been mentioned in the Introduction, are reducible to a process-driven 'positivist' model in which intersubjectivity, the communication event between two entities, becomes the ultimate yardstick with which to measure other communication processes. The embryo for this view is most commonly located in Shannon and Weaver's monograph *The Mathematical Theory of Communication* (1949).

As Chris Chesher (1997) appraises this text for its relevance to the Internet:

> Shannon and Weaver separated information content from the means of its carriage. ... While Shannon and Weaver's information theory was strictly technical and not concerned with the meaning of messages, it was applied far more broadly in describing and analysing communications processes. Information was imagined as a centre of social actions – the 'information revolution', and the 'information society'. The computer as information transferer and processor applied the epistemological model of 'information' in this tradition. Treating communications as largely a problem of 'getting the message across', and information as autonomous and central, has become a dominant and often uncriticised premise of contemporary info-culture. The operation relies on a belief that the informational sign unproblematically stands in for the actual. (88)

Shannon and Weaver's theory, a result of research conducted for the AT&T telecommunications company, aimed to account for how a unit of information which is produced by a sender at one end of a communication channel is able to be faithfully reproduced at the other end by a receiver. The source may be speech on the telephone, writing in a book, or beeps on a telegraph wire, which is conducted on a channel (a wire, a magazine or book) and received by another person with or without the aid of a 'decoding' device.[13] Such a sensibility about information has entered into everyday popular conceptions. The very nomenclature 'hi-fidelity' is erected on this model – that somehow an original performance of a musical piece can be faithfully reproduced on an electronic music system in one's living room. George Lakoff (1995: 116) points out that such a model also pervades the epistemology of education. The idea that teachers 'impart' knowledge to the minds of students which must then be 'regurgitated in an exam' presupposes that all knowledge is comprised of stable quanta of information, and that such information is understood by sender and receiver in exact duplication.

Shannon and Weaver's theory is pure medium theory: they were interested in neither the content of messages, their meaning, the possibility of intentionality behind them, nor the social and psychological condition of their reception. Yet their theory became a standard departure point for 'information theory' as it was appropriated by other disciplines and perspectives, including structural linguistics (particularly Roman Jakobson) and media effects theory. The distinction of this theory is that it rapidly claimed for itself a universal applicability, whether the kind of communication being examined was between machines, biological entities or human institutions.

It is not surprising that Shannon and Weaver's physics of communication could easily synchronize with the co-emerging field of cybernetics (Shannon was a student of Norbert Wiener – the heralded founder of this discipline). Wiener's *Cybernetics: or Control and Communication in the Animal and Machine* ([1948] 1961) appeared the year before Shannon and Weaver's text. In this text, perhaps one of the first formalized understandings of 'information' as an ontological force in social life was presented.

The combination of utopianism and anxiety which is expressed there seemed futuristic in the 1940s and indeed is in many ways typical of cyber-space literature today. For example, the current fascination with chaos and complexity theory is anticipated in Wiener's discussion of entropy as the tendency for system-based organization to deteriorate without constant management by ever greater quantities of information.

For the latter condition to prevail, a state of perfect knowledge and perfect exchange should exist in communication infrastructures. Wiener would probably be very satisfied with the open and unconstrained char-acter of computer-mediated communication on the Internet and Usenets. Together with the anthropologist Gregory Bateson, he promoted the circular realization of information flows containing feedback mechanisms, which for them was a necessary condition for communicative solidarity. Their critique of the mathematical theory was not a rejection of its posi-tivism as much as the fact that, as a model designed by and for commu-nications engineers, its unilinearity was unable to accommodate the social characteristics of communication processes.

Implicitly, of course, the schools influenced by cybernetics were at the same time critical of broadcast as an antisocial communication apparatus, consisting of unequal relations between senders and receivers, and a dis-tortion of information that resulted directly from broadcast's technical sub-structure rather than from class or ideological biases.

However, whilst the cybernetic schools may have rejected unidirec-tional modelling, the idea of feedback does not necessarily make the unilinear model of communication redundant, as John Fiske (1982) has pointed out:

> Feedback ... has one main function. It helps the communicator adjust his message to the needs and responses of the receiver. ... Though feedback inserts a return loop from destination to source, it does not destroy the linearity of the model. It is there to make the process of transmitting messages more efficient. (23)

So, by Fiske's account, the early cybernetic models added the fact that receivers were more actively a part of the communication process, but their role remained confined to a transmission model typical of the 'process schools'.[14]

It was not until George Gerbner's (1956) attempt at a general model of communication that the process school was able to break out of some of its more positivist underpinnings (i.e. that a medium is a transparent carrier of messages, and that the content of messages is objectively given, waiting to be faithfully reproduced).

In his model, the meaning of any given message is culturally relative as individual perceptions will order and make sense of a communication event in different ways according to the most familiar cultural frame-works available. The other major departure from the hypodermic model of communication is in postulating what 'a message' actually is. For Gerbner, a message never exists in some kind of raw state waiting to be coded, sent, then decoded. Rather, the practice of coding is itself part of what a message is. The revelation that accompanies this, however, is that the medium in which a message is sent is itself a part of the coding and therefore of the message – the means and control dimension of communication. The innovation which Gerbner makes therefore is in critiquing the idea that the medium or form of communication merely conveys, transports or transmits the message. Instead, the message is always already a part of the form.

In also making the listener, viewer or receiver more active in the process of communication, Gerbner introduces two new concepts: access and availability. The first refers to the social and technical conditions for access to a communication medium. In the second media age, not everyone can afford Internet access. First World ownership of television is high, *but* access to the transmission of messages is extremely low for most people. With the concept of availability, Gerbner points towards the closure of communication at the point of the production of a message. Before the age of mass media, the availability of 'information' was confined to relatively privileged or cloistered groups of intellectuals who had the literacy skills denied to the majority. In totalitarian political regimes, the population may be entirely literate but the central organization of power is based around the dissemination of selective publications, which has earned the title of propaganda. Here the selectivity and lack of availability of alternative literature, rather than what it says, is what makes it propaganda. Critics of propaganda seldom appreciate this fact, putting the influence of the material down to its 'highly charged' ideological character. Paradoxically, the same publication, when disseminated in democracies offering free speech, can be heralded as positive proof of this speech rather than derided, as it might be elsewhere.

Gerbner's insights, in taking the hypodermic model to extended lengths, offer some revelations about media 'form' to which we can return later. Nevertheless it should be pointed out that Gerbner still did not depart from the dyadic positions of the transmission models of communication.

The problem with positivist transmission models of communication is that they assume that all communication occurs in a vacuum without appreciation of the social and cultural contexts involved. For example, largely absent from transmission accounts is an appreciation that the 'success' of any particular communication depends on *the degree to which interlocutors might share a common culture.*

There are some *limited* exceptions to this in the models of Lasswell (1948) and Newcomb (1953). With Lasswell, the addressee is widened to include mass communication. Because of this fact, Lasswell's model has been a popular foundation paper for traditions in media studies, particularly the 'effects' tradition and audience studies. The fact that Lasswell insisted that mass communication needed a different methodological approach from personal communication makes his work useful for analysing broadcast. Lasswell was interested in the influence of communication structures on society as a whole. His most general and famous adage was: *Who says what, in which channel, to whom and with what effect?* Lasswell's framing of communication theory in this way proliferated into an array of sub-branches looking at content, control, audience and impact. But his guiding principles came from functionalist sociology, which recognized communication institutions as important in the regulation of social relations, and therefore in need of monitoring, improvement and policy so as to avoid 'dysfunction'. These principles address the role that communication processes can play in social reproduction. Mass communication, in particular, provides an inventory of public messages which allow social values to be monitored. In large-scale settings of social integration a media-generated consensus around social values enables better integration between society's institutions as well as maintenance of traditions and respect for the past.

Lasswell's work might be seen as articulating Durkheim's reference to communication, in the nineteenth century, as a material social fact which provides one of the ingredients of social solidarity and dynamic density: '… the number and nature of the elementary parts of which society is composed, the way they are arranged, the degree of coalescence they have attained, the distribution of population over the surface of the territory, the number and nature of channels of communication, the form of dwelling etc.' (Durkheim, 1982: 58). Like Durkheim, Lasswell also continued the nineteenth-century sociological dichotomy of society versus the individual in which communication is treated entirely as a social fact, that is, 'a category of fact with distinctive characteristics: it consists of ways of acting, thinking, and feeling, external to the individual, and endowed with a power of coercion, by reason of which they control him' (Durkheim, 1982: 52).

This dualism of society or 'system' versus the individual as the basic unit of the functionalist paradigm is successful to the degree to which 'media' are considered a continuation of social forms by technical means (see previous chapter), but it runs into difficulties when particular media are seen to be constitutive of new social forms (see the discussion of McLuhan below).

Whatever Lasswell's political aspirations as a reformer, his work has the merit of offering a general theory of communication that spans broadcast and network. Today the legacy of the Lasswellian approach, combined with the information thinkers, can be seen in the various discourses that try to grapple with CMC in the vast assortment of perspectives which are all nevertheless framed by process models: the user perspective, the content perspective, economic and political perspectives and control perspectives.

CMC as cyberspace   The benefit of the process models for studying the second media age is that they provide a departure from exclusively content- and linguistically based models of media analysis. In doing so, they begin to explore the 'ends of the chains' of communication events, taking into account the significance of who is speaking, the nature of the medium in which this speech occurs, and the effects of communication events for the listener.

However, the early information theorists are unable to address two important questions in CMC: the precise techno-social nature of the medium that 'mediates' in CMC, and the kind of identities that exist on-line.

To illustrate this, consider Gerbner's model. Gerbner's advance was to show how a sender's or receiver's appreciation of a medium could actually alter the content of an individual message to the point where, he argues, it is imperative that the medium-contexts of communication must always be taken into account. Of course, this insight is valuable if the medium that is implicit in the communication process is capable of reproducing the structure or appearance of an object or external reality (analogue communication). With digital communication, however, where there is no analogy entailed in the communication process, the ability of a communicant (who is virtually immersed) to make sense of what the digital substructure signifies socially is almost entirely lost. A prominent example is that of HTML, the mark-up language used for putting pages on the World Wide Web. When the pages are finished they can be analogically and graphically hyperlinked with other pages and interactively interfaced on screen. However, the mathematical code that underpins it plays little or no part in cognitive communication.

Interestingly, it is only when the complex binary code that underpins so much of what we actually see on the screen becomes rendered as an analogue interface that it begins to make sense – not as language, but as 'space'.

One of the central tenets of computer-mediated communication theory is that CMC enables a form of 'socially produced space' (Jones, 1995: 17), namely cyberspace. This is said to be comparable to a kind of electronic *agora*.[15] The *agora*, dating from post-Homeric Greece, refers to an open space in which goods and information are exchanged. In the *agora*, information is typically relayed by word of mouth or by messages posted on walls, a process which even became institutionalized in European life in the form of the cosmopolitan coffee house.

The café, which is frequently attributed with the status of bedrock of 'civil' society,[16] has of course become an extensive carrier of the proselytization of cyberspace with the large number of cyber-cafés that have sprung up in cities all over the world. These (embodied) cafés are places in which the rituals of the old world – coffee consumption – and of the new – logging on to an ICQ, MUD, MOO or email service – become entirely blended.

Paradoxically, these cafés, which are frequented by those interested in tapping into the civics of cyberspace, at the same time annul the embodied kind of civics for which cafés were originally invented. The café table is replaced by benches and rows of terminals appended to coffee-stained keyboards.

The other kind of café that is also intertwined with the advancement of CMC is the on-line virtual café, where, in a MUD or a MOO, partici-pants meet in an analogue representation of a café, present themselves to other café revellers, and engage in hours of chat.

According to Marc Smith (1995), there are four aspects of virtual interaction that shape the communication behaviours that go on within them.

- Virtual interaction is aspatial, whereby increasing distance does not affect the kind of interactions possible. Because of this, the economies of co-presence are superseded to the point where mutual presence becomes redundant in cases where it was once a functional imperative. Smith cites, for example, the growing trend for companies to relocate to rural areas.
- Virtual interaction via systems like the WELL is predominantly asyn-chronous. With the exceptions of Internet Relay Chat, MUDs and ICQs, CMC (e.g. conferencing systems and email) operates by the flexibility of posting messages which can be replied to according to the convenience of users' own time zone or work schedule.
- As with communities of scholars whose connection is mediated by print, CMC is acorporeal because it is primarily a text-only medium. The dual effect of the asynchronous and acorporeal features of CMC is its facilitation of interaction between quite large groups of people, well beyond, for example, what telephone conferencing could enable.
- CMC is astigmatic; that is, social differentiation based on stigma tends to be absent as there are few visible cues and markings or behaviours which locate an individual with a particular social status.

The last point here is one which Tim Jordan employs in his book *Cyberpower* (1999). For Jordan, CMC is inherently anti-hierarchical. He argues that because identity in cyberspace is seldom identified with the off-line hierarchies, differentiation based on status is very difficult (81). Secondly, the many-to-many capacity of the Internet creates a much more inclusive and participatory environment in which the culture of exclusion which occurs in off-line life is difficult to sustain.[17]

CMC and the problem of identity   Smith (1995) contends that the four characteristics of interaction that he advances combine to make virtual interaction reasonably anonymous. This, he claims, leads directly to issues of identity in a virtual space.

In many virtual spaces anonymity is complete. Participants may change their names at will and no record is kept connecting names with real-world identities. Such anonymity has been sought out by some participants in virtual interactions because of its potential to liberate one from existing or enforced identities. However, many systems, including the WELL, have found that complete anonymity leads to a lack of accountability. As a result, while all members of the WELL may alter a pseudonym that accompanies each contribution they make, their user id remains a constant and unambiguous link to their identity. However, even this fairly rigorous identification system has limitations. There is no guarantee that a person acting under a particular user id is in fact that person or is the kind of person they present themselves as. The ambiguity of identity has led some people to gender-switching, or to giving vent to aspects of their personality they would otherwise keep under wraps. Virtual sociopathy seems to strike a small but stable percentage of participants in virtual interaction. Nonetheless, identity does remain in a virtual space. Since the user id remains a constant in all interactions, people often come to invest certain expectations and evaluations in the user of that id. It is possible to develop status in a virtual community that works to prevent the participant from acting in disruptive ways lest their status be revoked. (www.netscan.sscuet.ucla.edu/csoc/)

This particular kind of anonymity which Smith describes as operating in CMC Jordan sees as a result of the fluidity of identity which operates in cyberspace. This fluidity, which is much more open than is possible in institutional, embodied life, necessitates individuals' creation of a stable self-identity. For this reason, CMC interlocutors tend to spend much more time than in other forms of communication revealing information about themselves, their status, place, and other contexts for why they are communicating. Jordan (1999) refers to such an identity as an 'avatar'.

An avatar is a stable identity that someone using Barlovian cyberspace has created. The existence of an avatar means someone has used some of cyberspace's resources in ways that result in other avatars recognising a stable online personality. Someone's avatar may be constructed from the style of their online writing, from the repeated use of a name or self-description, or from any number of other virtual possibilities. (59)

However, no avatar is ever stable for long, and its potential, if not actual, transiency is always working against its stability. Another feature of CMC which undermines this stability is the sheer mobility that it offers communicants. As Steven Jones (1995) has suggested:

The importance of CMC and its attendant social structures lies not only in interpretation and narrative, acts that can fix and structure, but in the sense of mobility with which one can move (narratively and otherwise) through the social space. Mobility has two meanings in this case. First it is clearly an ability to 'move' from place to place without having physically travelled. But second, it is also a mobility of status, class, social role and character. (17)

This mobility is highly evident in the phenomenon of the widening generation gap between adults and adolescents (see Holmes and Russell, 1999). The empowerment which adolescents experience by way of CMC immersion is intensified by the fact that it dramatically exaggerates the generation gap between them and pre-CMC generations. This gap rests on both the widening differentials in technical competence and the fact that many parents and teachers find CMC alien in the ways in which it promotes individual forms of adolescent self-construction. In addition, broadband interface technologies such as the Internet lead to rapid identifications with global concepts of citizenship. The cultural mores which emerge from the interface of adolescent and technology subsume the narrow rigidity which previously characterized family norms and conventional forms of discipline and pedagogy which exist within the classroom.

The new sense of the personal which emerges through CMC immersion establishes itself in differing ways. On the one hand, the investment of an adolescent's identity in avatars attenuates embodied or face-to-face relationships, whilst, on the other, it enhances the personal qualities of being an autonomous information consumer. Here the status of adolescents as by far the strongest take-up group of CITs becomes particularly heightened in the age of virtual communities.

Taking some cues off-line – contexts of CMC   The generation gap phenomenon that has been a feature of the take-up of CMC and CITs in general highlights an aspect of this perspective which has so far been overlooked. Whilst it is interesting to examine how the technical mediums of CMC may, to varying degrees, directly affect the forms of community and identity which operate within them, the outer contexts of CMC also need to be assessed. A prominent exponent of this view is Nancy Baym, who argues in her essay 'The Emergence of Community in Computer-Mediated Communication' (1995) that '[t]oo much work on CMC assumes that the computer is the sole influence of communicative outcomes' (139). This assumption is exemplified by what is called the 'cues filtered-out approach', which, she says, has come to dominate the understanding of computer-mediated communication:

> Because computer-mediated interactants are unable to see, hear, and feel one another they cannot use the usual contextualization cues conveyed by the appearance, nonverbal signals, and features of the physical context. With these cues to social context removed, the discourse is left in a social vacuum quite different from face-to-face interaction. (139–40)

Baym identifies five different sources of impact on CMC: external contexts in which the use of CMC is set; the temporal structure of the group; the infrastructure of the computer system; the purposes for which the CMC is used; and the characteristics of the group and its members (141).

With regard to the first source, Baym argues that '[a]ll interaction, including CMC, is simultaneously situated in multiple external contexts' (141). In other words, communication in CMC depends on the sharing of a common culture which antedates interlocutors exchanging meaning on the Internet. Secondly, like Marc Smith, Baym points out that interactions may be synchronous or asynchronous. The time-world of the exchange is in turn influenced by the sub-variables of the computer system infrastructure, the speed, number and adaptability of computer interfaces. Finally the number of interlocutors forming a group and the intensity of purpose of their interaction will significantly determine what is communicated and how. Baym's analysis is an important advance in overcoming the legacy of much information theory, which takes as its ideal-type dyadic reciprocity as well as transport notions of communication. These departure points are instructive up to a point, but have difficulty accommodating the complexity of CMC.

The convergence perspective

An important sub-variant of the second media age thesis, and one more encompassing than the relatively narrow concerns of on-line CMC perspectives, is found in the literature on media convergence (see Fidler, 1997; Flew, 2002; Van Dijk, 1999). The convergence paradigm can rest on an architectural distinction between broadcast and network, but sometimes also on an historical (second media age) distinction, as in the case of Van Dijk.[18]

Convergence perspectives range from looking at 'industry convergence', to medium convergence, to convergence of individual media technologies.

Technological convergence is the usual starting point for this perspective, and can take place at the level of infrastructure (transmission links – optical fibre, microwave, satellite) or transportation (content being transported in a new way such as Internet on TV, or webcasting). Services such as weather on phones, entertainment on the Internet, but also types of data, and the way sound, text, data and images can be combined, are all included under the umbrella of convergence. What underlies such convergence are various forms of the integration of telecommunications, data communications and mass communications (Van Dijk, 1999: 9).

There is also the 'functional convergence' that occurs in individual media products, such as mobile phones converging with digital cameras. This sense of convergence is perhaps the most commonplace, and mostly takes the form of pointing out how older 'analogue technologies' have been re-created in the image of digital technologies. There are also very crude attempts to suggest that television and the Internet exhibit some kind of essential process of merging because TV can be viewed on computers, and CMC is readily advertised on television (see Seiter, 1999: 115).

But these technological levels of convergence are only made possible by industry convergence, resulting from collaboration between corporations in telecommunications, media and IT, or by takeovers and mergers between them. The relationship between corporate and technological convergence is dynamic and two-way. Corporate convergence gives rise to new combinations of mediums, technological innovation and content delivery, whilst technological innovation creates the compulsion for new kinds of corporate convergence.[19]

But finally, we can speak of the convergence between broadcast and networking as *mediums*, which Van Dijk (1999) calls 'the second communications revolution'. This revolution is one in which older mediums are redetermined, in two key ways – as interactive and as digital. It is digitalization which, according to Flew (2002: 10–11), is significant for the way in which it makes platforms and their media inter-operable, and networkable. Moreover, digitalization delivers the cybernetic dream of separating a channel of communication from content. Digital media can be broken down to a common base of bits, which are universally transferable and manipulable across media.

So what converges then, in terms of mediums, is not digital and analogue technologies, but new digital technologies, with digitally remediated analogue technology, as Table 3.1 outlines.

Under the broadcast column, television, radio and newsprint are each available in a digital form, as are many of the consumer items associated with them, such as DVDs and personal computer portals for viewing or listening to such media. Notable is a return to a wired infrastructure as providing a wider bandwidth for broadcast media, and the decline of electromagnetic transmission, which otherwise require an analogue-to-digital conversion process for use by the end consumer in digital form.

In the network column, there are also older analogue network technologies, most obviously the telephone, which was one of the first to be digitalized, in landline exchanges, but also older analogue mobile networks. However, unlike the broadcast column, there is also an array of 'born digital' technologies, which have been made possible entirely within a network infrastructure context. The Internet is at the frontier of these technologies, but the digital telephone network is also the hub for a proliferation of new P2P (person-to-person) networked bandwidth.

On the policy front, broadcasters are interested in the 'free speech' implications of a second media age, and exploit the way in which its historicism has become an orthodoxy by lobbying government regulators to slacken ownership concentration laws. Meanwhile key players in network media who facilitate the de-commodification of broadcast products, software, music and film-downloading web portals, are attacked by the owners of such media products via civic-legal means or by relayed pressure through telecommunications authorities. In this case, however, the arguments are not political, concerned with freedom of speech, but exclusively economic.

Table 3.1 Digitalization as the basis of convergence, wider bandwidth and multi-media (the ability to combine image, sound and text)

	Broadcast (wireless and wired)	Network (wireless and wired)
Technology	Digital TV, radio on-line, news text on-line, DVDs	Older network technology: 'digital-enabled' ISDN, mobile telephony, from analogue to digital
		Altogether newly 'born digital' technologies: the Internet, mobile text, mobile fax, mobile data, mobile video text, networked PDAs (Personal Digital Assistants). Providing new services: home shopping, banking, gambling, searchable databases
Medium-channels	Electromagnetic waves, satellite, microwave, copper and optical fibre cable	Satellite, microwave, copper and optical fibre cable
Policy	Broadcasters pressure governments to relax policies, because 'everyone' can be a broadcaster	Networking makes possible the provision of more information and entertainment that is otherwise commodified by broadcasters and telcos (telecommunications corporations) and now provided for free. MP3, movies, news – dilutes the user-pays dimension of media
		Advertising not as powerful, but its promise has caused losses for broadcasters

For convergence theorists, technologies, media and policies have each become more interdependent across both broadcast and network architectures of communication:

> No medium today, and certainly no single media event, seems to do its cultural work in isolation from other media, any more than it works in isolation from other social and economic forces. What is new about new media comes from the particular ways they refashion older media and the ways in which older media refashion themselves to answer the challenges of new media. (Bolter and Grusin, 1999: 15)

The virtual urbanization perspective

To confine the discussion of convergence to technology and industry can overlook the ways in which very profound convergences are happening between electronic and physical spaces. Compared to the research on media convergence, there is much more literature dealing with the convergence between technology and urban life.

Whereas both convergence and the CMC perspective have largely neglected the influence of external realities, as Baym (1995) has pointed out, there is certainly a growing body of literature which can be and which is distinctively concerned with the urban contexts of media cultures, both new and old (see, e.g., Boyer, 1996; Crang et al., 1998; Droege, 1997; Graham, 2000; Graham and Marvin, 1996; Hall and Brotchi, 1991; Hepworth and Ducatel, 1992; Holmes, 1997; Mitchell, 1996; Ostwald, 1997; Soja, 1996; Swyngedouw, 1993; Turow and Kavanaugh, 2003). This literature, which is oriented towards a 'virtual urbanization' perspective, is beginning to explore the multiple interrelations between computer-mediated space and contemporary urban space. As I have argued elsewhere (Holmes, 1997), virtual realities are already embodied in particular everyday technologies such as freeways, television and the shopping mall (41–2). The most instructive feature of these technological forms is that they can already be considered as proto-virtual realities, displaying features that characterize virtual spaces: they tend to homogenize 'culture' according to a logic which makes experience over in its own image; and they presuppose and are major contributors to the production of cosmopolitan world-spaces.

Implicitly, virtual urbanization is a perspective which sees life on the screen not so much as a new or additional development but as an internal development of the logic of modernity and of the kind of urbanization that accompanies it.

Whereas, with conceptualizations of the first media age, broadcast is viewed as a neutral kind of service to urbanization and city life, it can now, with the Internet, be seen much more as a precondition of this urbanization. Globally, urban populations are increasing much more rapidly than the rate of increase of overall world population. According to UN estimates, urban dwelling worldwide was 39% in 1975, 50% in 2000, and will be 63% in 2025.

The central features of urbanization that are relevant to the rise of the second media age are:

- the increased scales of spatial separation of workplace from household;
- the atomization of an urban population into units of consumption;
- the standardization of the built environment;
- the privatization and duplication of access to property and resources.

Urbanization, but also suburbanization, is enhanced by an increase in transportation and communication technologies which have traditionally enabled the maintenance of geographic association. As I have argued elsewhere (Holmes, 2001: 31), '… as cities grow in their scale and abstractness, the older technologies of urban connection, the motor car, the television, the telephone, can become inadequate for the maintenance of a daily cycle of connection'. Because of this, new forms of communicative networks such as the Internet and mobile telephony are lauded for their perceived speed and efficiency in being able to replace the relatively cumbersome networks of old. Alternatively, there is also a tendency to abandon urban networks altogether, as they become replaced by the security village, the gated community, within which the home becomes the ultimate refuge. The retreat to ever-shrinking spaces of urban privacy – the home, the graphics of the personal computer, the intimacy of the SMS keypad – leads to ever greater 'personalization'. This in turn opens up markets for the commodification of the means to ever greater control over immediate environments – be this with the car, the home cinema or electronic assistance that is carried on the body.

The more individuals find connection to an 'outside' on the basis of personalized domestic electronic refuges, the less need there is for the main street or the *agora*, and for public space in general. What public space remains tends to itself be characterized by very enclosed and privately controlled environments.

Both the physical and the electronic urban architectures converge around the principle of continuous subdivision. Such subdivision is endless in its scale, as the connectivity that these network spaces make possible enables ever-expanding forms of urbanization – cities on ever greater scales. Within these scales, each domestic or individual unit is separated and united at the same time. The relationship is dialectical: the more the individual cell is able to be integrated via a centralized, decentralized or distributed network, the less dependence there is on proximity to or physical immediacy with others. Conversely, the less individuals get to know their neighbours, the more important it is to be tuned in or logged on to one of these networks.

These relations of synthesis through segregation are already anticipated in Debord's account of the social bond via the image and of Adorno and Horkheimer's depiction of the culture industry, which is further discussed in the next chapter. However, what they do not analyse is the way in which media forms coalesce with other urban means of connection.

For example, an important dominating virtual space is the contemporary freeway. The technical as well as metaphorical links between the motor vehicle superhighway and the so-called information superhighway are quite profound (see Jones, 1995: 10–11). Freeways and electronic media are deeply embedded in the suburbanization process as well as making possible ever greater separation between workplace and home and leisure zones – such as the mega shopping malls.

As I have argued elsewhere (Holmes, 2004), the path-dependence on motorized transport and the path-dependence on telecommunication are mutually constitutive. When our social world becomes geographically fragmented, we privately come to rely on the agents of separation that have aggregately produced this condition. Wherever freeways have driven corridors of speed and efficiency through a city, it creates a culture of avoiding accidental contact with strangers. Emphasis is placed on the control which individuals have over their use of urban space. As we become more monadic in retreating to spaces from which we can exercise this control, such virtualization has the added 'benefit' of removing us from physical danger from embodied others, at the same time as it encourages us to fear others. As soon as we leave our car we become a delinquent, as Baudrillard quips about the LA freeway system. The motor car is as much an agent of protection as it is of transport. Our fleeting association with others travelling at high speed is unfulfilling in any physical sense of exchange as other drivers become our objects of 'road rage' just as anonymous interlocutors on the Internet can subject each other to 'flaming'.

The return of McLuhan

Having addressed recent literature on the urban and technical dynamics of cybersociety, it is instructive to return to the work of Marshall McLuhan as a case example of a thinker who provides a very early account of network media culture.

What we can nominate as McLuhan's 'second media age', which he calls the age of 'automation' or cybernation, is contrasted with the mechanical age of mass reproduction, which is the first media age. However, in McLuhan's texts we can identify two prior forms of media-tagged societies in relation to which the mechanical/electric distinction operates. These are 'tribal' social conditions based on speech and scribal society based on alphabetic writing. Together, the four kinds of society – tribal, scribal, mechanical and electric – do not evolve in a linear progression, but rather each kind of society can encompass a number of qualities which are found in others. Moreover, McLuhan does not posit an over-arching process to the development of these revolutions. It is only in the electric age that what he calls the sense ratio and sensory balance that individuals have with their environments becomes stable once again. This results in what he views as a 're-tribalization' of culture, a return to days of audile sensory stability, before the distorted technological mediums of writing and print.

The mechanical age is characterized by fragmentation but uniformity, repetition and centralism corresponding to the first media age, whilst the electric age is one of integration via decentralization, which creates 'extreme interdependence on a global scale'. To a large degree, individuals in information societies are still catching up with the new possibilities of technically constituted worlds because they are caught off guard by the biases of print culture. Certain adjustments in psycho-social life are necessary before we can face 'electromagnetic technology'.

For McLuhan, information, rather than vision, becomes the basis of the electric age. In a passage very similar to Jean-François Lyotard's claims in *The Postmodern Condition* (1984: 194), he argues that the era of cybernation is one in which prior forms of technological extension will not be allowed to exist except by being translated into information systems (McLuhan, 1967: 68; see also Innis, 1972).

A well-known distinction that McLuhan makes which roughly corresponds to a first versus second media age thesis is that between 'hot' and 'cool' medium. Hot mediums include radio, movies, photographs. Cool mediums include the telephone and TV.

> A hot medium is one that extends one single sense in 'high definition'. High definition is the state of being well filled with data. A photograph is, visually, 'high definition'. A cartoon is 'low definition', simply because very little visual information is provided. Telephone is a cool medium, or one of low definition, because the ear is given a meager amount of information. And speech is a cool medium of low definition, because so little is given and so much has to be filled in by the listener. On the other hand, hot media do not leave much to be filled in or completed by the audience. Hot media, are, therefore, low in participation, and cool media are high in participation or completion by the audience. (McLuhan, 1964: 31)

These relationships can be represented as in Table 3.2. From this table it can be observed that McLuhan partially subscribed to an 'informational' view of communication, in which senders and receivers become connected by a message. The receiver of hot messages may have quite a bit of work to do – depending on the medium. Notably also, McLuhan does not distinguish between technologies of broadcast (like TV) and point-to-point network technologies (like the telephone).

There are a number of difficulties with McLuhan's explication of hot and cool mediums, however – like his rather strained distinction between cinema and television. Firstly, it is true that cinema is able to provide 'more information' than television, especially if a sense-impression view is taken[20] – it has a wider screen – but the ability of television to convey complexity is quite outstanding compared to other forms of media. Secondly, McLuhan claims that hot mediums tend to extend only *one* sense in high definition. His classification of cinema as 'hot' is glaringly out of place in this regard. Thirdly, McLuhan claims that a cinema or radio audience is passive whilst a television one is more active. During his own time of writing and to date, no empirical audience studies have shown this to be true. Fourthly, McLuhan contradicts himself where he says that hot mediums tend to overtake cool mediums, but that, historically, it is 'past mechanical times' that can be designated as hot' whilst the 'contemporary

Table 3.2 Features and types of hot and cool mediums

	Hot mediums	Cool mediums
Features (McLuhan, 1964: 31–3)	Low participation	High participation
	Extends one single sense	
	High definition	Low definition
	A large amount of information	Small amount of information
		Need to be completed by the audience
	Tend to ovetake cool mediums	Tend to be supplanted and remade by hot mediums
	Tend to be mechanical, repetitive, uniform	
Examples (McLuhan, 1964: 31–2)	Radio	Telephone
	Cinema	Television
	Photograph	Cartoon
		Speech
	Phonetic alphabet	Hieroglyphic and ideogrammatic written characters
	Print	Monastic and clerical script
	Paper	Stone
	Lecture	Seminar
	Book	Dialogue
Examples (McLuhan, 1964: 36)	Past mechanical times	The contemporary TV age
	Developed countries	Underdeveloped countries

TV' age is cool (36). Finally, McLuhan describes how cool mediums 'need to be completed by an audience'. However, the actual instances of cool mediums he specifies – telephone, cartoon, speech (if taken as face-to-face), the interactive seminar and dialogue – don't actually have an 'audience', whereas the hot mediums of radio, cinema, print and lectures do.

For McLuhan's schema to have coherence, television should be placed in the category of a hot medium. If this were done, his account would make sense in terms of the first and second media age – corresponding to what he calls the mechanical age versus cybernation. I point this out, not in order to fit everything into the dualism that this distinction purveys, but because McLuhan himself is inconsistent.

A final problem is the way the hot/cool distinction is over-extended to include all manner of objects, past and present societies, stone and paper, phonetics and writing, so much so that it dilutes itself by way of a generalized and generalizing dualistic vitalism.

However, in differentiating between hot and cool media according to definition and information, different technologies are clustered in a parallelism around like mediums according to broadcast and network integration. That is, the telephone (cool), even though it is interactive, allowing for significant 'participation' in a communicative event, is merely a conduit for speech, but lacks the (hot) range of cues and information involved in the face-to-face. Television, one would think is, in itself pretty 'hot', but compared to its 'counterpart', cinema, with its imposing high-quality screen and sound, television is actually quite cool. Christopher Horrocks (2001) claims that the Internet has problematized the status of television as a cool medium. Cool mediums demand high participation, and wide bandwidth interactivity is a heightened form of participation otherwise denied to television. Horrocks suggests that because the Internet acts like a kind of meta-medium in combining a range of media, of which television becomes part of its content, it renders television not so 'cool' anymore and the latter can be considered 'hot' in relation to the Internet, if we allow it to be considered as a distinct medium.

McLuhan's distinction between hot and cold media places the capacity for a medium to convey complexity at a premium – in the volume of information (what the cyberspace theorists would see as bandwidth), the level of definition (e.g. the continuums that McLuhan never really wrote about – analogue to digital) and the number of senses that it works upon. Notably, McLuhan's observations on hot and cool media are a remarkable anticipation of the theoretical analysis of 'virtual reality'.

As Nguyen and Alexander (1996) point out, for McLuhan, 'every new technology changes how our sense organs operate to perceive reality, and it may be that computer technology changes not only our perception of reality but also our very selves' (112–13).

> Media, by altering the environment, evoke in us unique ratios of sense perceptions. The extension of any one sense alters the way we think and act – the way we perceive the world. When these ratios change, [individuals] change. (McLuhan and Fiore, 1967: 41)

However, notably, we are never aware of the hold which a medium has over us when it is effective and operating. As soon as we notice a medium, it becomes 'old'. This is arguably why it is possible to speak of a 'second media age' now that the Internet and network-interactive technologies are more prominent – the older mediums of broadcast are rendered more object-like, rather than invisible.

Social implications

Cyberspace as a new public sphere

One of the most prominent implications of the purported 're-tribalization' of the consequences of the second media age is the way in which it is said to allow a renewal of the public sphere. During the 1970s a number of thinkers heralded the decline of the public individual and of the public sphere (see Gouldner, 1976; Habermas, [1962] 1989, and, later, 1974; Sennett, 1978). Post-broadcast accounts of the public sphere lay claim to new kinds of politics, new kinds of 'electronic assembly', and even a return of participatory democracy by way of CMC. In the 1990s the most enthusiastic promoters of electronic democracy came from the editors of *Wired* magazine. Jon Katz prophesied the emergence of a 'digital nation' in which on-line culture would offer the means for individuals to have a genuine say in the decisions that affect their lives, whereas Kevin Kelly saw in the Internet a revival of 'Thomas Jefferson's 200-year-old dream of thinking individuals self-actualizing a democracy' (cited in Lax, 2000: 160).

In a key text addressing the role of the Internet in transforming the nature of the public sphere, Mark Poster (1997) claims that 'contemporary social relations seem to be devoid of a basic level of interactive practice' (217). For Poster, the physical forums for 'interactive practice ... such as the agora, the New England town hall, the village church, the coffee house, the tavern, the public square, a convenient barn, a union hall, a park, a factory lunchroom, and even a street corner' (217), are in decline. The central factor behind such a demise of embodied assembly is, according to Poster, the concomitant rise of broadcast media which 'isolate citizens from one another and substitute themselves for older spaces of politics' (217).

Poster takes up John Hartley's argument that, for all intents and purposes, broadcast media *are* the public sphere: 'Television, popular newspapers, magazines and photography, the popular media of the modern period, are the public domain, the place where and the means by which the public is created and has its being' (Hartley, 1992b: 1). In Hartley's view, the media provide a specular space which, although it lacks the possibility of direct interaction, allows participants to express public opinion through the act of consuming media as well as to relate to a common culture of discourses. If it is true that, as Hartley would suggest, the electronic media have eclipsed and displaced the public sphere, then a great deal of pressure is placed on understanding what kind of public sphere electronic media produce.

The specular space of the media, where all participants can relate to message producers and the messages that are produced, is one which, up to a point, sits well with Jürgen Habermas's idea of an homogeneous universal public sphere. In a central work, *The Structural Transformation of the Public Sphere* ([1962] 1989), Habermas defines the public sphere as a domain of uncoerced conversation directed exclusively towards pragmatic agreement. For Habermas (1989), such a development of a democratic public sphere was possible in the seventeenth and eighteenth centuries but has been diluted in the current period by the fact that the apparatus of media is controlled by interests which systematically distort the content of public discourse.[21]

However, what is also stressed in Habermas's earlier work is the importance of 'literacy' in the formation of discursive publics. For him, the press was at the centre of a rational project towards democracy. Taking Britain as a model, Habermas argues that capitalist entrepreneurs promoted the 'world of letters': 'The public sphere in the political realm evolved from the public sphere in the world of letters' (1989: 30–1). Through the salon, theatre, and the coffee house, 'Conversation turned criticism and *bon mots* into arguments' (31) as public discourse became autonomous from church and court. Looking for emancipation from church and state, the rising bourgeoisie appealed to enlightenment values of 'free speech' and debate in the same stride as they sought to remove the obstacles to a free market.[22] Such values enabled educated and propertied classes to maintain ideological power, but nevertheless upheld the ethos of freedom of opportunity and the sense of citizenship that accompanies this.

The extent to which the Internet and new 'interactive' technologies facilitate and maintain the literacy necessary for Habermas's rational project is pivotal here in deciding what contribution they can make to any form of democratic deliberation. Certainly, studies of how Internet submedia are used show that they are highly text-based, but to what extent is such textual communication merely a reproduction of off-line communication? And to what extent do personal computers using graphic user interface share with TV and video games the privileging of 'emotion and empathy instead of reason and judgement'? (Kaplan, 2000: 208).

In a volume looking at the idea of global literacy on the Web, Hawisher and Selfe (2000) ask: 'How does the ordered space of the Web affect the literacy practices of individuals from different cultures – and the constitution of their identities – personal, national, cultural, ethnic – through language? What literacy values characterize communications practices in this ordered space?' (1)

They critique the claims of Net ideologists such as MIT Media Lab director Nicholas Negroponte and former US Vice-President Al Gore that the Web is a culturally neutral literacy environment. Such a claim is derivative of an imperializing, 'global village' narrative which 'is shaped by American and Western cultural interests at the level of ideological production' (1).

The ethnocentric ideology of the global village heroically imagines the information networks which the West supplies to 'the world' as some kind of paternalistic gift-of-community. Or, as Hawisher and Selfe put it rather more cynically:

> According to this utopian and ethnocentric narrative, sophisticated computer networks – manufactured by far-sighted scientists and engineers educated within democratic and highly technological cultures – will serve to connect the world's peoples in a vast global community that transcends current geopolitical borders. Linked through this electronic community the peoples

of the world will discover and communicate about their common concerns, needs and interests using the culturally neutral medium of computer-based communication. When individuals within the global community discover – through increased communication – their shared interests and common-weal, they will resolve their differences and identify ways of solving global problems that extend beyond the confining boundaries of nation states. (2)

What makes a lie of such a narrative is the fact that the greater majority of the world's population does not have access to the Net or World Wide Web. Rather, 'the culturally specific nature of literacy practices clearly influences the use of the Web and the use of the internet in fundamental ways' (2), especially when we consider what Selfe calls the 'ideologically interested nature of the global-village narrative as constructed specifically within the framework of American and western politics and economics, and culture' (2).

The fragmentation of the bourgeois public sphere

If we take Hawisher's, Selfe's and Habermas's observation that the bour-geois public sphere is very much confined to the educated and literate classes, and, globally, their concentration is in powerful Western nations, it becomes difficult to conflate an Internet-mediated public sphere with anything like a 'global village'. Whilst every nation and every population is part of the globe, not everyone partakes of this idyllic public arena.

However, not even the bourgeois public sphere, in its limited form, is as unitary as cyber-utopians claim it to be. Since Habermas put forward his thesis of a unitary public sphere, many theorists have suggested alter-native models, such as Negt and Kluge's (1993) idea of an 'oppositional' working-class public sphere, Rita Felski's concept of a feminist public sphere, and Nancy Fraser's (1990) notion of a 'post-bourgeois' public sphere.

What is distinctive about these last mentioned models is that they each define themselves against a unified public sphere as pervaded by some version or other of a 'dominant ideology' – be it patriarchal or bour-geois or perhaps 'logocentric', and based too much on decision-making and questions of 'consciousness'. More recently, newer understandings of a public sphere have emerged which can be viewed as qualitatively different from traditional civic and media-extended accounts of 'publicness'. These newer theses take account of interactive media and 'interactivity' as considerations in the delimitation of alternative possibilities of civic integration.

Todd Gitlin (1998) has advanced the idea of 'public sphericules', seg-mented spheres of assimilation which have their own dynamics and forms of constitution. Gitlin argues that 'a single public sphere is unnec-essary as long as segments constitute their own deliberative assemblies' (173).

Gitlin suggests that the segmented assemblies constituted by computer-mediated communities do loosely interrelate, in a parallel sphere of liberal-pluralist diversity. It is akin to the state-generated public sphere implied by 'multi-culturalism', in which a citizen in, say, Australia or America might adopt a national identity by embracing a much more unitary principle of publicness – liberal or communitarian pluralism.

Finally, the segmentation of the public sphere comes to bear down on the question of democracy itself:

> Does democracy require a public or publics? A public sphere or separate public sphericules? Does the proliferation of the latter, the comfort in which they can be cultivated, damage the prospects for the former? Does it not look as though the public sphere, in falling, had shattered into a scatter of globules, like mercury? The diffusion of interactive technology surely enriches the possibilities for a plurality of publics – for the development of distinct groups organized around affinity and interest. What is not clear is that the proliferation and lubrication of publics contributes to the creation of *a* public – an active democratic encounter of citizens who reach across their social and ideological differences to establish a common agenda of concern and to debate rival approaches. (173)

Gitlin does not address the role of CMC in traditional kinds of decision-making activities like voting, which characterize participatory kinds of democracy (see Sobchack, 1996), but, rather, suggests that the electronic public sphere, what John B. Thompson (1995) calls 'mediated publicness', facilitates a 'deliberative' model of democratic engagement.

Gitlin's view accords with the thesis of Barbara Becker and Josef Wehner (1998), who argue that interactive media support the formation of 'partial publics' – 'discourses characterised by context-specific argumentation strategies and special themes' (1).

Becker and Wehner still subscribe to the idea that traditional mass media have the central role of mobilizing and institutionalizing public opinion, but argue that interactive media are growing in significance as a space for the formation of 'pre-institutional' forms of public opinion.

Interactive media enable alternative kinds of public opinion, but this 'alternativeness' does not come out of ideological reaction to dominant values in the media, but from the structure of interactive mediums themselves. Thus, Becker and Wehner follow Neidhardt and Gerhards in arguing that different forums of public opinion – based on direct or extended interaction, on assemblies, or on the mass media – correspond to different ways of 'selecting, clustering and spreading information' (Becker and Wehner, 1998: 2).

Technologically extended interactive environments are distinguished from mass media by the fact that they are unable to constitute a 'mass' in which individuals are related together as 'citizens'. Rather, the Internet promotes differentiation instead of homogenization by 'generating poly-contextual communication structures' in which there 'is no citizen who is discussing with other citizens on the net. Rather, there are simply individuals – such as experts, old people, homosexuals, women, men, children, youngsters – who debate their particular interests on the net' (Becker and Wehner, 1998: 2). Becker and Wehner echo many of the advances made by the second media age theorists. However, they add two important observations which challenge the characterization of the Internet as a free de-centralized structure by firstly pointing out that the numerous sub-media of the Internet are characterized by 'thematically restricted domains' – a point to which I will return. Secondly, less and less information on the Net can be regarded as 'public' and universally accessible as, increasingly, the bulk of Internet content becomes colonized by contextless, fragmented information (advertising, spam, unverified messages) whilst a significant volume of 'bandwidth' is accessible only by institutional and private elites.

Public/private

What both the models of 'unified' and 'partial' politics discussed above are committed to is some notion of the separation of the public from the private, which rests on the Greek distinction between *polis* (the place of *demos* – democracy) and the *oikos* (household.) The public/private distinction is a complex one, which in modern capitalism is so often confused by the extension of private control (private property, private interests) into the 'public sphere' as market place. The traditional pre-capitalist market place is not a place of private interests negotiating but of the public good of exchange. Today the private exists in the public sphere, as can, to take Hartley's argument, the 'public' exist in the private. Privacy might be commonly thought of as being confined to the spaces of the home, but this is also, increasingly, the place where, paradoxically, individuals gain access to the public sphere.

This is mutually generated; the less individuals engage in practices of interaction in 'public spaces', the more they are likely to be engaged in interactive practices in private spaces, and vice versa. Under these conditions the household unit becomes a primary cell of modern social relations, the basic unit and building block from which social interaction occurs. When the public sphere has withdrawn to the home, where a 'dialogic' or two-way open interaction becomes impossible, interaction becomes more and more 'confined' to the family, the household and one's workplace.

These conditions certainly did not obtain in pre-media society, in which the frequency and intensity of embodied interaction were of an entirely different order. The origins of European modernity since the eighteenth century, for example, are founded on the café as the bedrock of the emergence of a public sphere (Habermas, 1989). In the year 1700, for example, the city of London boasted 3,000 coffee houses.

However, in media societies, where the geographic and kinship ties of the parish, local neighbourhood or industrial slum have virtually disappeared, individuals have historically become very heavily dependent on media of many kinds to acquire a sense of belonging and attachment to others. The situation is one of separation and unity. Individuals are separated at a geographic level, locked away in their housing-allotment or -unit fortresses, but united on scales of city or nation in their attachment to forms of media. Ironically, the marketing calls for consumers to 'get connected' and 'travel on the Internet' instead of being 'stuck at home' are an exact reproduction of the social and urban consequences of broadcast technologies. Individuals are told they can interact to overcome the tyranny and restraints of broadcast, but they do so only by reinforcing the domestic conditions of their atomized existence.

The question of whether interaction, once it is *reduced* to the electronically mediated and technologically extended kinds of access to communication which are enabled from the home, constitutes participation in a public sphere is a pivotal one to ask in relation to CMC. Certainly the private/public question becomes extremely vexed on the Internet. As Poster (1997) suggests:

> If 'public' discourse exists as pixels on screens generated at remote locations by individuals one has never met and probably will never meet, as it is in the case of the Internet with its 'virtual communities', 'electronic cafés', bulletin boards, e-mail, computer conferencing and even video conferencing, then how is it to be distinguished from 'private' letters, printface and so forth? (219–20)

We could add to Poster's observation the fact that virtual meeting places are replicated in physical form in cybercafés and video-cafés. Symbolically as well as functionally, the cybercafé is extremely interesting. It strongly re-affirms the idea that the cellular network basis of gaining access to the public sphere predominates, where even one of the strongest institutions of embodied public life can be remade in terms of CMC. Nobody meets face-to-face at a cybercafé, as the face-to-screen interaction precludes dialogic contact in any form other than the electronic.

Problems with the public cybersphere thesis

The success of any argument claiming a special role for the Internet in the constitution of a new public sphere rests on its ability to establish a practical/imaginary unity in which all participants have equal opportunity for 'observation' and communication. This postulated imaginary unity, best known in the phrase 'virtual community', seldom reconciles itself with the fact that the Internet is not at all technically homogeneous and is segmented into quite a range of properties and capabilities which each carry different sociological and communicative potentials and effects.

It is true that, unlike television, the Internet is a network[23] as well as 'dialogical', capable of a two-way dialogue. But its network properties are rarely realized in communication directly, and seldom do they become *meaningful qua* network, because, as Becker and Wehner (1998) point out, individuals only ever 'use' the Internet within well-defined sub-mediums.

Trevor Barr (2000) usefully breaks down the different kinds of inter-action on the Internet into six categories:

1 one-to-one messaging (such as email);
2 one-to-many messaging (such as 'listserv');
3 distributed message databases (such as USENET news groups);
4 real-time communication (such as 'Internet Relay Chat');
5 real-time remote computer utilization (such as 'telnet'); and
6 remote information retrieval (such as 'ftp', 'gopher' and the World Wide Web'). (118)

It can be seen from this list that the Internet provides a generic environ-ment for a number of different modes of interaction which can vary according to real time/stored time, symmetrical versus asymmetrical dialogue, broadcast sending and receiving, and information posting and retrieval.

But each of these modes of interaction relates very differently to the possible constitution of an 'electronic public sphere'. Moreover, the infor-mation and communication possibilities of the Internet are more often than not parasitic of broadcast-mediated communication. The growth of companion websites which accompany media organizations, news-papers, consumer products, sporting events, etc., has provided an aston-ishing impetus to the use of information retrieval, listserv and interactive databases available on the Internet.

When CMC is broken down into specific sub-media rather than reduced to the indeterminacy of the Internet as a communication envi-ronment, a more sophisticated appreciation of the technological transfor-mations of the public sphere is enabled, and the advancement of new accounts of context-specific partial publics is one outcome.

However, at the same time, the global reach and mobility of all forms of Internet communication, regardless of the specificities of their sub-media, also need to be accounted for. The reason for this, I argue, is that it is impossible to separate the significance of contemporary CMC from its antecedent and wider context of broadcast communication culture.

Why this is significant is that, whereas broadcast generates an instant 'international context' of social connection, there are few ways in which individuals can achieve meaningful *interaction* to make tangible these global connections. There are telephones and other 'narrow-band' ways of communicating, but none of these is quite able to provide a multi-media context for any given interaction. The Internet, it is argued by its promoters, changes all of that.

In accounting for the growth of computer-mediated communication via the Internet, both national and global statistics become significant. But given that the experience of community on the Internet is not limited to national boundaries, it is also important to consider the shape and structure of this virtual community.

Besides being hailed as a technology which can deliver the 'global village', the Internet is also promoted as a singular medium which allows for democratized processes which were not previously possible in the era of broadcast. But what kinds of democracy are being postulated here? Traditionally, and more than ever now, democracy is heavily aligned with the nation-state (see Hirst and Thompson, 1996). Because of this, a nonsense is made of the claim that the Internet enables universal participation in the democratic process. The point here is that practices of communication afforded by CMC may be able to substitute some of the functions of the mass media – for example, in the formation of pre-institutional public opinion – but do not necessarily exert pressure on the institutional apparatuses of politics. Of course the mass media themselves, as a means of electronically mediated communication, can never replace the institutional apparatuses of politics, and, as numerous studies have shown, have been just as much used by politicians as they have influenced them.

The Internet can properly be classified as a 'global' technology, which enables connections with individuals and institutions overseas just as easily as it does nationally, regionally or locally. If there is an imagined community (see Anderson, 1983) on the Internet, it is definitely *not* the nation-state. State-bounded kinds of citizenship cannot be considered coterminous with the kinds of citizenship which are achieved on the Internet. However, this is not to argue that a global sense of citizenship, even if it too is an 'imagined one', cannot exist. Recent protests against international financial institutions such as the World Bank were organized almost entirely through Internet media – a case of not so visible electronic assemblies producing very visible embodied assemblies.

Democracy and interaction

To privilege either 'broadcast' or interactive mediums like CMC as domains which can deliver a universal public sphere is fraught with methodological problems. Perspectives on media epochs – 'the video age', the 'age of the Internet' (Turkle) or the 'second media age' (Poster) – are too simplistic and read as much too technologically determinist insofar as they neglect the sub-media and subcultures which are internal to apparatuses of electronic media, both broadcast and interactive. Such models tend to be one-dimensional in that they view forms of public association, be they by images and broadcast or by information and interactivity, as mutually exclusive.

At the same time, however, the 'public sphericules' or 'partial publics' thesis of Gitlin, and of Becker and Wehner, purveys another kind of technological determinism, which moves from the grand historical grounding of social life on one or other over-arching technology, to differentiating forms of association in specialized 'spheres' on the basis of more particular technological mediums as the context for particular civic subdivisions.

It is true that certain mediums, particularly ones like CMC which enable global reach, provide the individual with mobilities of communication which enable associations beyond what persists in modern life as the most powerful sense of a pre-given public frame – the nation-state. And moving beyond the nation-state in a global rather than 'international' sense also expands the numbers of those with whom we would want to participate in a public sphere. However, it is also true that individuals are mobile across communicative mediums and continuously participate not in a pre-given public sphere, but in the process of constructing publicness across a range of mediums. But it is less accurate to say that the contemporary public sphere is breaking down and becoming fragmented than it is to say that it is sustained across increasingly more complex, dynamic and global kinds of communication environments.

Notes

Parts of the discussion of the public sphere in this chapter, were presented in Holmes, D. (2000) 'Technological Transformations of the Public sphere: The Role of CMC', 2nd international conference on cultural attitudes towards technology and communication, Perth, July.

1 The mission statement, which is published at NetAid's website (*http://www.netaid.org/netaid/mission.htm*), is revealing in its appeals to a humanistic universalism.
2 For an analysis of the significance of the Walkman from a cultural studies perspective, see Du Gay et al. (1997).
3 Because of this, personalized information technologies are sometimes described as primitive or proto-virtual realities (Holmes, 1997). They are kinds of virtual realities, but ones which only seal off a restricted number of senses.
4 As Heilig points out, the human eye has a vertical span of 15 degrees as well as a horizontal one of 180 degrees (quoted in Shields, 1996: 76).
5 For a useful, pithy chronology of the development of the Internet, see Hobbes Internet Timeline at *http://www.isoc.org/guest/zakon/Internet/History/HIT.html*.
6 I will not rehearse here the numerous accounts of the military origins of the Internet's technical scope – that is, as an unintended consequence of the US military-industrial complex's need to decentralize information to avoid possible loss or capture of command and control information during projected nuclear exchanges (see Rheingold, 1994: 774).
7 The face-to-face is re-created *metaphorically*, by way of the promise of more comprehensive technical means to represent human gestures and communication. This at least is the promise that it offers to a mythologized dyadic relationship. An enhancement of intersubjective presence supplies its motivating ideology whilst, in its actual operation as a *system* of interchange, material displacement of such a relation is its outcome.
8 The 'guarantee' of reciprocity which once found its conditions in bodily present interchange and agency-extended reciprocity (reciprocity that does not pretend to be re-creating

the face-to-face but recognizes its instrumental, tool-using extension) becomes more and more displaced as social ties are more intensely recast in terms of dependency on technologically mediated simulations.

9   See also Robin Nelson's (1999) comparison of a TV drama from the 1980s and one from the 1990s, 'Boys from the Blackstuff 'and' Twin Peaks', respectively, as a case study in the way 'TV space has begun to depart from a (humanist) depiction of characters and events grounded in a historical world in which actions have consequences to a (post-modern) collage of attractive but dislocated images and sounds' (17).

10   It could be argued that the sums paid to celebrity sports and film stars are an accurate reflection of the commodification of the value of interactivity that is extracted from the media-constituted 'masses'.

11   A useful guide to decorum on the Internet can be found at *http://www.dtcc.edu/cs/rfc1855. html*; see also Shea (1994).

12   For a site which explores emoticons, see *http://www.randomhouse.com/features/davebarry/emoticon.html*

13   Shannon and Weaver's theory was actually motivated by the perceived need to enable cost-effective communication which minimized random noise and so allowed transparent and successful communication.

14   The codification of communication theory into process schools and semiotic schools is implicit in the work of John Fiske. This clustering of schools is potentially useful for the first and second media age distinction. However, semiotics is barely of interest for analysing the Internet, which is arguably all process and not much in the way of content. To the extent that it *is* content, there is no particular reason why it is any more instructive for the purposes of study than the 'wider' bandwidth semiotic mediums of broadcast.

15   For an argument for the case that cyberspace cannot be said to constitute a space in any socially meaningful sense, see Chesher (1997).

16   Indeed the German coffee house of the eighteenth century is heralded as no less than the place in which the communicative culture of modernity began in Europe.

17   For an argument which is strongly opposed to Jordan's and Smith's position, see Spears and Lea (1994).

18   Van Dijk (1999) restates the archetypal second media age thesis: 'The interrelationship of processes and the growing role of media networks gives rise to a new type of society. The best name for this new type is "network society". In the course of the twentieth century it has replaced another type of society that has been called "mass society"' (23).

19   A salient example is the marriage of digital photography and mobile telephony, leading to mergers between corporations who specialize in these technologies (e.g. Sony Ericsson), which in turn lead to innovations in the micronization of photographic technologies and the way images can be sent wirelessly.

20   Mark Poster (1990) claims that McLuhan's 'mediumization' of media studies is restricted to a sense-centred empiricist conception of media effects.

21   Electronic media become arms of the interests of capital rather than information providers which abide by a public service ethos. This ethos only remains in other institutions like public libraries, museums and government statistical services which are involved in impartial and neutral presentation of information.

22   Similarly, Hawisher and Selfe argue (see discussion below) that in the USA the popular imagination about computers is that they are the 'latest technological invention in a long line of discoveries that will contribute to making the world a better place by extending the reach and the control of humankind, most specifically the reach of America and its related system of free market capitalism and democracy' (Hawisher and Selfe, 2000: 6).

23   Television can be a network, but only for the content producers.

FOUR

THE INTERRELATION BETWEEN BROADCAST AND NETWORK COMMUNICATION

Thus far, we have looked at broadcast media and network media as distinct fields of enquiry for contemporary communication theory. As foreshadowed in Chapter 1, there are two principal ways in which this can be done: by taking up the 'media age' thesis, or by seeing broadcast and network forms of communication and association as making possible distinct forms of social integration.

In this chapter we will attempt to theorize the interrelation between these two forms, both in terms of the 'first and second media age' and in terms of 'social architectures' of media form. In both models, the way in which individuals find connection with the different media forms can be shown to be interdependent – network communication becomes meaningful because of broadcast, and broadcast becomes meaningful in the context of network. But the oscillation between these forms is not an entirely new phenomenon, and, as we shall see, predated the arrival of a 'second media age' by many years. It is just that in contemporary times this dynamic has visibly attained much more of a 'technological' and commodified separation.

As suggested in Chapter 1, my critique of cyber-utopianism is whether an historical distinction between the first and second media age can be made at all. I am strongly in agreement with the idea that broadcast and network communication mediums offer different possibilities of connectedness in information societies and that to contrast them is highly instructive, but to say that the latter has eclipsed the former is extravagant. Rather, I am going to argue, when other social contexts, such as the urban realities of their consumption, are considered, the two are mutually constitutive, and the mutuality of these dynamics was evident long before the Internet, as we shall see.

The first and second media age as mutually constitutive

As previously argued, the historical distinction between the first and second media age is the primary foundation upon which utopian claims about New Media have been made. This distinction underpins the second media age thesis itself and much of the cyberculture literature which now defines itself in opposition to, or as having succeeded, 'media studies'.

Part, if not most, of the difficulty in making this historical distinction between these two forms of media association is that it is based on a misconceived 'parallelism' between communication mediums and the technical mediums they are said to relate to as if in a one-to-one correspondence. Broadcast can be interactive as much as interactivity can be facilitated within broadcast. In fact almost all technically constituted forms of communication, from print to television, to cyberspace, contain elements of broadcast and interactivity; it is just that these are realized differently, and at different levels of embodiment in different techno-social relations.

In the historical model, broadcast media are characterized by one-way communication. Typically, this entails a sender of messages transmitting information to an indeterminate mass or audience, without that audience having recourse to also 'transmit' information, at least not to the extent that the broadcaster does. Post-broadcast mediums of communication, on the other hand, are said to provide for two-way interaction and a 'restoration' of the specificity of both interlocutors. In this way, the second media age is viewed as redemptive and emancipatory. The centrist tyranny that is seen to be carried by the apparatus of broadcast media is said to be annulled by a supposed democratization of broadcast, where everyone can be a broadcaster or datacaster, thereby flattening out the otherwise concentrated (performative) power of broadcast.

The overriding evidence for this argument is to point to the massive take-up of new media in developed nations. Statistics on the rate of growth of web traffic, the take-up of PCs in the home, as well as connections to the Internet, mobile telephony and short messaging services (SMS) or texting are all a part of this evidence. Regardless of what individuals actually *do* with the technology, or what it might *mean* to use it, the fact of its take-up is said to be proof of the need individuals have to 'find connection in a computerized world' (Rheingold, 1994).

It is empirically true that, from 1990 onwards, the take-up of interactive media technology in information societies increased far more dramatically than did the adoption of new broadcast technologies. However, if such a trend is posited as the basis of the second media age thesis, there are a number of problems.

One such problem involves the question of historical determination. The early second media age advocates like Gilder, Negroponte, Kapor and Poster suggested that the need for interactive technology has been historically created by broadcast. The very development of New Media which provide such interactivity is, in a sense, seen to be driven by this need. Second media age advocates suggest that the new interactive media are able to overcome the hard-wired asymmetry of broadcast and allow everyone to be a broadcaster and audience member simultaneously.

Moreover, culturally, the 'control society' age of broadcast is said to be fast disappearing. Audiences will not tolerate such subjection when they can be producers. According to the theory, they will rapidly abandon broadcast as a source of information and entertainment. Or if they remain loyal to broadcast, they choose reality TV, where they can see themselves in the production, rather than obey the mass-produced artifices of the culture industry.

However, from the historical vantage point of over a decade of the Internet, we can see how, empirically, this has not proved to be overwhelmingly true. Internet use in many of the most information-rich countries with high media densities began to slow in 1999. At the same time, as various studies indicate, attachment to broadcast forms of media did not show any significant decline (see, e.g., Castells, 2001; Schultz, 2000: 208). Furthermore, it is empirically the case that, when websites begin to charge fees for information that is otherwise free to air, net users rapidly abandon them.

The fact that the Internet as a communicative technology has not signalled a demise of radio, TV, newspapers or other broadcast media in information societies is tied to the fact that broadcast and network technologies are, as we shall see, mutually constitutive. Moreover, it reveals the fallacy of the technological determinism inherent in the first and second media age distinction and the way in which the heralding of a second media age is often represented as a linear eclipse of the broadcast era. At the core of the distinction between first and second media age is the idea that an historical era, defined by its media, can so closely correspond to a small number of technological forms. In the context of the current distinction, the fallacy inheres in reducing 'interactivity' and 'broadcast' to a function of technology itself.

As was argued in Chapter 1, neither broadcast nor interactivity needs to be technologically extended in order for its distinctive political, social and economic properties to be realized. For example, reciprocal communication is inherent to a range of technological forms, from face-to-face, to telephone and writing. From different perspectives, numerous surveys of the history of communications show how broadcast has had a systemic form which is as old as human society itself (see Feather, 2000; Innis, 1972; Jowett, 1981; Thompson, 1995; Williams, 1974, 1981). For Raymond Williams (1974), the social basis of broadcasting long preceded its mechanical and electronic forms:

> The true basis of this system had preceded the developments in technology. Then as now there was a major, indeed dominant, area of social communication, by word of mouth, within every kind of social group. In addition, then as now, there were specific institutions of that kind of communication which involves or is predicated on social teaching and control: churches, school, assemblies and proclamations, direction in places of work. All these interacted with forms of communication within the family. (21)

Garth Jowett (1981) has charted how, long before the development of writing, personal seals, portraits on coins and printed illustrations provided a system of broadcasting symbolic imagery to denote political, financial or intellectual authority.

For Williams (1981), the only difference between modern and pre-modern systems of broadcasting is that, today, audiences assemble in numerous combinations, in contrast to the 'massing' of previous kinds of audiences:

> ... we have at once to notice that there are radical differences between, for example, the very large television audience – millions of people watching a single programme, but mainly in small unconnected groups in family homes; the very large cinema public – but in audiences of varying sizes, in public places, on a string of occasions; and the very large actual crowds, at certain kinds of event, who are indeed (but only in this case) physically massed. (15)[1]

Broadcast and network interactivity as forms of communicative solidarity

Whilst the historical basis for making a distinction between broadcast and interactivity is weak, the sociological basis for making this distinction is strong. The repeated observation made by second media age theorists that the take-up of interactive technology is a way of overcoming broadcast is, I argue, an important one. It suggests that, in media societies, the horizontal integration of direct two-way interaction provides aspects of social integration, and a sense of belonging that cannot be provided by the 'vertical' kinds of integration of broadcast media.

However, the second media age theorists offer few answers as to why this yearning for interaction is a yearning for *technologically extended* interaction, and cannot be satisfied by face-to-face interaction.

Since the formation of publics mediated by broadcast apparatuses, face-to-face interaction, and its extended forms, which may be synchronous (in low bandwidth tele-mediated interaction such as telephone) or asynchronous (such as letter writing), have provided horizontal interaction in ways that complement broadcast interaction. This fact can be seen in numerous studies over the years which demonstrate how vertical and horizontal kinds of interaction have historically been co-extensive.[2]

The fact that broadcast and interactivity operate mutually is most visible at the level of content:

- The programming material of broadcast media – the soap narratives, the sporting events, the personality of media presenters, the content of the news – provides the content of countless conversations, be they face-to-face, on the Internet, or as other kinds of interaction. The fact that the Internet is parasitic on broadcast can be found in what is actually discussed in computer-mediated communication environments. For example, in the mid-1990s the largest subscriber list of all on Usenet (over 3 million users) was r.a.t.s, which stands for rec.arts.tv.soaps (Baym, 1995: 138). Baym (1995) claims that '[t]he idea of soap opera fans using computers to gossip about their favorite (and least favorite) soap characters challenges conventional images of both soap opera fans and computer users' (138).[3] Mathew Hills (2001) argues that on-line fan clubs function as a community of imagination that is always 'available' to fans:

> As an affective space in which caring for, and about, the object of fandom constitutes the most significant communal claim, the fan newsgroup resembles an ongoing and never-ending fan convention. … It is thus a 'space' in which common sentiment can migrate from a fixed or ritualistic point, moving out into the fan's practice of everyday life via the newsgroup's constant availability. (157)

• Equally, however, the content of broadcast media is overwhelmingly composed of simulations of, or references to, face-to-face interaction. Consider the soap opera, the most popular TV genre: it is made of up of thousands upon thousands of face-to-face interactions, or images of people on the telephone, and, more recently, the Internet. But mostly they indulge in never-ending close-up shots of dyads and triads of faces interacting. We watch the TV, 'hooked' on our programme, as if the faces, gestures and expressions which stream past the screen give us a simulated dose of the exact same kinds of interaction we are avoiding ourselves – as long as we remain a viewer. Similarly, radio genres engage all of the metaphorics of face-to-face company which listeners are suspending as a condition of that engagement. The radio announcer thanks us for 'joining the programme', or declares it is 'good to be with you'. All of these statements are to suggest the announcer is in the room with you. But it need not be the face-to-face that is substituted in this way. It is not insignificant that advertisers use telephone rings, or doorbell or alarm clock sounds, with increasing frequency to also get the listener's attention.

Each of these examples shows how broadcast media are a central reference for face-to-face interaction, and extended forms of interaction, as much as this interaction itself becomes the content of broadcast genres. However, in the context of the second media age thesis, this co-dependent relationship has become obscured. Instead the second media age thinkers only emphasize one side of this relationship, that of the need for individuals to renew interactive forms of communication in relation to the way broadcast media displace geographic forms of interaction.

Broadcast and interactive communication also operate mutually because they have central characteristics common to them, which are each aspects of the commodification of communication. This manifests itself in four ways: (1) as technologies of urbanization, (2) as technologies of mobile privatization, (3) by enabling forms of the extension of social relationships, and (4) as agents of interrelated economic processes. When these continuities are recognized, many of the social and political claims of the second media age perspective are shaken.

Communication and urbanization

Both broadcast and interactive forms of communication structure, and are structured by, one common operation – the partitioning of a mass into atomized units. In this a number of observations of the second media age theorists about the first media age are most noteworthy. What they describe as the first media age is charged with the characteristic of interpellation and individuation.

Adorno and Horkheimer (1993) were among the first to provide an insightful description of this effect of media in their discussion of the 'culture industry'. For them, the culture industry corrodes the horizontal networks of associations which make up urban self-formation, channelling populations into individual units, who, once isolated, must reconnect through what vertical means is offered to them via the mass media. 'City housing projects designed to perpetuate the individual as a supposedly independent unit in a small hygienic dwelling make him all the more subservient to his adversary' (30), which for Adorno and Horkheimer is the power of capitalism to remove individuals from networked means of cultural production.

For them this process is circular: the more individuals become reliant on media, the less they are dependent on horizontal networks. The more abandoned are these networks, the more mass media become their only source of cultural production. Thus, the entire edifice of mass media becomes an environment of what has more recently been called 'path dependence'. In turn, the need for people to form local attachments to place is removed, as place is redefined as anywhere within a common mediascape. The 'need' for everyone to stake out their own self-enclosed unit, preferably on a greenfield site on the fringes of suburban expansion, rather than adjacent to dwindling inner-urban horizontal networks, is a frontier expression of media-driven urbanization.

However, this kind of urbanization is usually explained, first, in terms of redefining individuals as consumers and, secondly, by assigning them 'identical needs, in innumerable places to be satisfied with identical goods'. The organizers of media industries declare such a culture system to be 'based in the first place on consumer's needs, and for that reason ... the technical contrast between the few production centres and the large number of widely dispersed consumption points ... [is, they claim] accepted with so little resistance' (Adorno and Horkheimer, 1993: 31).

Adorno and Horkheimer's point is that mass media both produce media products to satisfy consumer needs as well as produce these needs to satisfy the culture system of the media. In other words, there is a sense in which media audiences themselves are turned into commodities – as consumers who have a use-value for the growth of the culture industry itself: 'The result is the circle of manipulation and retroactive need in which the unity of the system grows ever stronger' (31).

Thus, media-driven urbanization, which requires the duplication of innumerable, individual units wherever it operates, breaks up networks at the same time as it installs a more over-arching form of assimilation around a 'few production centres' of culture. In doing so it even menaces the extended network possibilities of the telephone. 'The step from the telephone to the radio has clearly distinguished roles. The former still allowed the subject to play the role of subject. … The latter … turns all participants into listeners and authoritatively subjects them to broadcast programmes' (31).

Given their predilection for the role function that is offered by the telephone over radiated media, Adorno and Horkheimer might be expected to welcome the liberal qualities of the Internet, and the ability of consumers to interact with previously one-way forms of programming. However, from an economic and social point of view, the introduction of a more powerful interactive apparatus in the context of the dominance of mass media is of an entirely different order than the transition from telephone to radio and television.

Once the social path-dependence on the architectures of radiated media is in place, interactivity, no matter what form, must be conducted on the basis of such architecture, and indeed contributes to the daily replication of such an architecture. This is something which second media age advocates fail to point out in their claims of the redemptiveness of cyberspace: horizontal unity overcomes the vertical segregation, but also reproduces this segregation.[4] Moreover, also neglected by the second media age theorists is that the first media age is characterized by a form of separation and unity. As we saw in Chapter 2, there are many agents through which this occurs, the image being primary among them – the society of the spectacle.

With Debord (1977), spectacle is a form of reification, a realm in which direct social relationships are expressed through a totemic system of images. To the extent that the spectacle 'concentrates all gazes and all consciousnesses' (aphorism 3), the individual only 'recognizes himself in the dominant images of need' (30). The channelling of attention towards a singular medium is the very basis for segregated individualism. There is no competing need for horizontal gazes and dialogue, which the spectacle accommodates entirely. Within Debord's terms, however, it is not *essential* that the image becomes the agent of separation/unity; it is just that in the most intense period of high capitalism's self-promotion it is a 'technique' which serves such a role. The important point is that in late capitalism social needs require abstract 'mediation' in some form:

If the spectacle, taken in the limited sense of 'mass media' which are its most glaring superficial manifestation, seems to invade society as mere equipment, this equipment is in no way neutral but is the very means suited to its total self-movement. If the social needs of the epoch in which such techniques are developed can only be satisfied by their mediation, if the administration of this society and all contact among men can no longer take place except through the intermediary of this power of instantaneous communication, it is because this communication is essentially unilateral. (Debord, 1977: 24)

To the extent that the Internet can be considered a medium that is at once instantaneous and an invocation of the gaze (the World Wide Wait), it signals little change in the 'individuating' aspect of the technological mediation of embodied presence. In other words, if an 'audience' is constituted only in an atomized form by mass media, then the difference between the phenomenological world of a broadcast audience member and that of an individual immersed in a so-called 'interactive technology' begins to flatten out.

Some empirically driven research on Internet use confirms the acceleration of individualization which typifies CMC. The Stanford 'Internet and Society Study' conducted by the Institute for the Quantitative Study of Society (Nie and Erdring, 2000) found that Internet users spend more hours at the office and keep working when they get home, and the longer people have used the Internet, the more hours they spend on it per week. As the director of the Stanford study, Norman Nie, explains:

We're moving from a world in which you know all your neighbors, see all your friends, interact with lots of different people every day, to a functional world, where interaction takes place at a distance ... the more hours people use the Internet, the less time they spend with real human beings. (Nie and Erding, 2000: 1)

The use of Internet sub-media brings individuals together at the level of electronic assembly but it also renews the physical atomization of media operators. In doing so it materially creates the very conditions which ideologically it proposes to overcome.

To this degree, network communication is actually parasitic of one of the conditions that have been produced by broadcast whilst continuing this condition. The need for extended network communication is proportionately related to the degree of geographic atomization which exists within a communicative field.

Broadcast can be considered a first media age in relation to the fact that its atomistic qualities seem to be more tangibly overcome by the Internet. But in truth, they are also 'virtually' overcome by the medium of broadcast also. The concept of a first media age begins, therefore, to look more like a theoretical invention integral to the postulation of a second media age.

Communication and mobile privatization

The role of communication in urban cultures cannot simply be reduced to the multiplication of urban units but relates to the way the communicative mediums which they make possible allow for ever greater mobilities from within these units. This tendency is accentuated by the diffusion of new media, as the virtual urban futures thesis maintains (see Chapter 3), but in truth it is characteristic of all electronic media, not just those of a second media age. As I have argued elsewhere (Holmes, 2001), electronic media do not have to be portable to provide mobility, as it is possible to 'travel' with audio-visual and interface media, even when the body is in stasis.

The products of information capitalism are themselves commodities, but their most significant quality is that they commodify subjective and intersubjective experience. In the context of urban cultures, the endgame of such a process culminates in private bubbles of communication and commuting.

The definition of 'private' is here extended from its typical use in economics and politics. As Margaret Morse (1998) uses it in a spatial sense, privatization refers to the way in which certain urban spaces, like privately owned and controlled shopping malls and the home, become refuges which de-realize their 'outside' environments. For example, the historical separation of workplace and household becomes exaggerated by broadcast media, where the shrinking public realm only has enough room in it for work itself, whilst the pursuit of leisure is expected to take place in the home. 'The process of distancing the worker from the workplace and the enclosure of domestic life in the home, separated from its social surroundings, allowed a compensatory realm of fantasy to flourish' (109). For Morse, the home physically disconnects itself from an outside, virtualizing itself on scales that expand as the public realm contracts.

Such privacy, described in the information technology industry as 'personalization', was advanced by Raymond Williams (1983) in his concept of mobile privatization: 'At most active social levels people are increasingly living as private small-family units or, disrupting even that, as private and deliberately self-enclosed individuals, while at the same time there is an unprecedented mobility of such restricted privacies' (188). Williams argues that the private 'shells' of the motor car, office and home unit gradually become extended by new media in ways in which it becomes possible to travel without physical movement. The paradox of MTV or the World Wide Web is that thousands of images can stream past us every hour, where we can be transported around the world at lightning speed, sampling countless other places, styles and impressions, whilst we are stationed in absolute stasis, our only motion being with a mouse or remote control. New media give us a mobility which exempts the consumer from having to leave the comfort of his or her shell, even his or her armchair. Such shells do not have to be physical; indeed, the space in front of a computer screen is one that is intensely personalized, designed for a single user with the right password, and with icons, characters and images which demand a face-to-screen association.

Mobile privatization is not merely a trend in domestic and civic life; it also mediates the modern workplace (see Greenfield, 1999) and post-workplace trends evident in telework (see Morelli, 2001). The most important feature of Williams' concept for the present discussion is that media-based mobility can be interactive or derived from broadcast. In *Television, Technology and Cultural Form* (1974) Williams describes how the increasingly private refuge of home and family generates the need for 'new kinds of contact' and 'news from "outside", from otherwise inaccessible sources' (27). Such a need generated from the fact of urbanization may be met just as much by letters and telephone as it is by Internet connections, mobile telephones and texting. However, the more mobile and portable such means of connection are, the more we can translate the private refuge of the home into the electronically generated recluse of communication bubbles which de-realize our relation to physical public space. The more such bubbles are occupied, the less likely are face-to-face forms of recognition. What face-to-face relations do exist in the public sphere are jeopardized, or made more schizophrenic, by the always open possibility that a mobile phone will ring or a message device will beep.

The endgame of mobile privatization is, according to Arthur and Marilouise Kroker (1996), a radically divided self – a self which is at war with itself – split between an embodied self and an electronic identity. Because of this, there is a demand on each individual to find protective shells in which to co-exist. The Walkman and the mobile phone are powerful examples of the micro-personalization of the self in public life – what the Krokers call the 'electronic self':

> ... the electronic self is torn between contradictory impulses towards privacy and the public, the natural self and the social self, private imagination and electronic fantasy. The price for reconciling the divided self by sacrificing one side of the electronic personality is severe. If it abandons private identity and actually becomes media (Cineplex mind, IMAX imagination, MTV chat, CNN nerves), the electronic self will suffer terminal repression. However, if it seals itself off from public life by retreating to an electronic cell in the suburbs or a computer condo in the city, it quickly falls into an irreal world of electronic MOO-room fun within the armoured windows. (74)

The electronic self seeks to 'immunize itself against the worst effects of public life' by 'bunkering in'.

> Bunkering in is the epochal consciousness of technological society in its most mature phase. McLuhan called it the 'cool personality' typical of the TV age, others have spoken of 'cocooning' away the 90s, but we would say

that bunkering in is about something really simple: being sick of others and trying to shelter the beleaguered self in a techno-bubble. (75)

According to the Krokers, such an electronic self becomes incapable of negotiating a pre-virtual public world and becomes politically tuned out and socially awkward:

> Suffering electronic amnesia on the public and its multiple viewpoints, going private means that the electronic self will not be in a position to maximize its interests by struggling in an increasingly competitive economic field. (74)

Denied a political conduit, the electronic self is reactive only – often swaying between fear and anger:

> Frightened by the accelerating speed of technological change, distressed by the loss of disposable income, worried about a future without jobs, and angry at the government, the electronic self oscillates between fear and rage. Rather than objectify its anger in a critical analysis of the public situation, diagnosing, for example, the deep relationship between the rise of the technological class and the loss of jobs, the electronic self is taught by the media elite to turn the 'self' into a form of self-contempt. (76)

Bunkering in, which becomes a kind of survival strategy, tends to narrow the electronic self's world of control to 'whining about the petty inconveniences' rather than broad scales of social and political issues. 'Bunkering in knows no ethics other than immediate self-gratification' because its outerworldly contact has been de-realized. This disengagement created by the architecture of the processed digital world makes it 'hip to be dumb, and smart to be turned off and tuned out' (76).

> The psychological war zone of bunkering in and dumbing down is the actual cultural context out of which emerges technological euphoria. Digital reality is perfect. It provides the bunker self with immediate, universal access to a global community without people: electronic communication without social contact, being digital without being human, going on-line without leaving the safety of the electronic bunker. The bunker self takes to the Internet like a pixel to a screen because the information superhighway is the biggest theme park in the world: more than 170 countries. And it's perfect too for dumbing down. Privileging information while exterminating meaning, surfing without engagement, digital reality provides a new virtual playing-field for tuning out and turning off. (77)

In the Krokers' depiction, those features that are attributed to the TV age – cocooning, the bunker ego, the passive couch potato, etc. – find themselves multiplied in the context of the greater array of electronic bunkers which provide a new platform of engagement/disengagement in social life.

Technological extension

> Rapidly, we approach the final phase of the extensions of man – the technological simulation of consciousness, when the creative process of knowing will be collectively and corporeally extended to the whole of human society, much as we have already extended our senses and our nerves by the various media. (McLuhan, 1964: 11)

Any form of interaction which exceeds embodied mutual presence requires the extension of one or more communicative faculties. A number of the thinkers examined in the previous chapter view media in terms of an overdevelopment of one or more aspects of sensorial culture – print, the image, oral culture – so much so that, in the case of McLuhan, it is proposed that an entire epoch can be based on it.

However, the property of extension, which may enable the continuation of face-to-face kinds of cognitive communication across time and space, also introduces entirely new qualities which are not possible within face-to-face communication. And it is *this* property of extension that is common to both broadcast and interactive technology.[5] Any medium which enables such extension will necessarily transform the content, form and possibility of what can be communicated. As Meyrowitz (1995) suggests, analyses of mediums are significant because they suggest that 'media are not simply channels for conveying information between two or more environments, but rather shapers of new environments themselves' (51). For this reason, writing and print can also be included as technologies of extension. Both are able to dispense with presence insofar as they can reach across both space and time (see also Sharp, 1993: 232).

Technological extension, which can never reproduce the fact of embodiment (despite the ideological project of virtual reality), nonetheless facilitates a great number of possibilities which mutual presence cannot fulfil. Two of its broadest features are that technological extension can transform:

- the symmetry of a communication process (as with broadcast);
- the temporal mode of presence of a communication process (as with delayed interactivity versus real-time communication and storage retrieval – synchronous or asynchronous).

When it is technologically extended, broadcast, which is possible within Internet sub-media like email and bulletin boards as well as traditionally regulated apparatuses, typically involves a temporal separation between the point of production of communication and the point of consumption. For example, television has been described as 'a vast relay and retrieval system for audiovisual material of uncertain origin and date which can be served up instantaneously by satellite and cable as well as broadcast transmission and videocassette' (Morse, 1998: 107). The production and consumption of information are typically separated in context and in time, a separation which can be unlimited in its degree of abstractness. On the Internet, the six kinds of interaction described by Barr (2000) in the previous chapter (see p. 79) nearly all involve asynchronous forms of mediated connection.

However, the Internet is unique in its ability to combine possibilities of synchronous and asynchronous engagement – by extending the properties of speech, writing and the image in combination. The Internet is itself a storage network as well as an interactive environment. Electronic mail may be dyadic and reciprocal, circular within discussion groups, of a broadcast nature broadcasting information, with any particular user transmitting to the many. Powerful search processes are able to retrieve information stored and collated in databanks. The only difference between television and the Internet on these points is that the 'retrieval' of content from their respective archives is performed by different people: by media workers on behalf of an audience (in the case of television), or directly by the consumer (in the case of the Internet).

The economic interrelationship

A fourth continuity between the broadcast and network architectures is the way their economic logics presuppose each other. Economically, network media are accompanied by a form of commodification which, as we shall see, constitutes a parasitic reversal of the kind of commodification which typifies broadcast. The form of commodification which is peculiar to a first media age is intertwined with the ever-expanding dimensions of advertising.[6] As Smythe (1981) shows, television, and any kind of broadcast media, do not so much sell products as sell the concentration spans of audiences to advertisers. In this relationship the circulation of the sign becomes fused with the circulation of commodities:

1 Advertisers sell material commodities to the consumer.
2 Broadcasters sell audiences (consumer consciousness) to advertisers (measured by ratings).
3 Television programming is marketed by broadcasters to consumers.
4 Consumers are also workers who sell their labour-power to corporations in return for the means to purchase commodities.
5 The cost of advertising is reflected in the price of the commodity.

For step 2 to occur, broadcasters have to be able to offer high numbers of media consumers and high-quality 'concentration', which is achieved during 'prime time' when higher prices for advertisements are charged.[7] This simply cannot be achieved in any medium other than broadcast, which is why, as I have outlined elsewhere (Holmes, 1997), advertising will never be successful, in a 'stand-alone' way, on the Internet as it is in mass media. One of the reasons for the crash in dot.com stocks in 1999 was the failure of advertising on the Net as a revenue-generating exercise (see Cassidy, 2002; Kuo, 2001). On the Internet, there is no mass, because a mass is entirely constituted by broadcast media. Broadcast audiences are built up over time, but they need to be reasonably synchronous with regular visibility for advertisers to have any success.[8]

The interactive sub-media of the Internet, like the World Wide Web, cannot deliver a ready-made 'mass' to advertisers, and even when websites are used in a companion role, their effectiveness is questionable. Gauntlett (2000) points out that 'modestly sized but inescapable adverts' still appear on the World Wide Web and these ads can be personalized to particular sites, but the solvency of web directories is overwhelmingly derived from the promise of stock values, not from advertising revenue (7). Chan-Olmsted (2000) observes that 'the Internet is the most cluttered medium in the world. To succeed in marketing an online brand, a marketer most likely will need distribution of communication messages via mass media to create broad awareness of the product or service' (98). Direct marketing ads are nearly always duplications of advertising 'that has come from channels outside the Internet, such as TV spots and infomercials' (98). Internet sub-media that allow consumers to personalize information they want will only ever be a 'valuable extension' of traditional media' (100).

Thus, broadcast advertisements often include a World Wide Web address, so that 'the consumer can continue a brand relationship initiated in an ad in an established medium and extend it to a closer relationship on the Net' (100). The fact that such sites might get a large number of visits is already driven by mass marketing. However, web proprietors (portal providers and search engine companies) have vigorously tried to aggregate web advertising in terms of portal loyalty strategies. In what has been called a game of 'portalopoly', 'Internet companies are racing to build sites (known as portals) that serve as hubs or gateways to the larger internet' (Buzzard, 2003: 205). Portals function like the mass circulation magazines or TV networks: 'They are sites that meta-aggregate content and offer a range of services in order to be the home page for as many users as possible, thereby attracting more advertising revenue' (Buzzard, 2003: 205). However, as we shall see below, contra Buzzard (2003: 206), these hubs, lacking liveness, performativity and specular visibility, fail to provide a mechanism for audience constitution.

On the Internet, the technical nature of its sub-media means it is not possible to be a broadcaster in the same way as it is with *mass* media. Yet, it is on the basis of a deception, or at least a sociologically uninformed view, that broadcasters in the USA are lobbying to have ownership and control laws relaxed in the wake of the Internet (cf. McChesney, 2000). Because anyone can publish content on the World Wide Web, they argue that media monopolies have been rendered ineffectual and so restrictions on owning television, radio and print outlets should be lifted.

Media of interactive communication, however, tend to be scarce in advertising and rest on the time-charging of a communicative event, renting of telephone lines, renting of computer servers and storage space. Conversely, the more broadcast media are provided 'free-to-air', the more they feature advertising. Viewers are relatively tolerant of such advertising, but the same does not obtain for forms of broadcast that require pay-per-view, such as cable TV and cinema. Consumers of pay-television and time-charged pay-per-view media resent advertising in these contexts. Cinema-goers arrive late at films in order to avoid lead advertisements, whilst cable TV must promote itself as advertisement-free.[9] An example of the indignation at receiving advertisements via a pay-per-use service surfaced in Australia in August 2001, when the largest phone company, Telstra, was investigated over charging customers for 'spam' messages. The Australian Consumers Association demanded that all customers should get a refund. A spokesperson for the Association, Charles Britain, said: 'I think it's a bit rich. It's a characteristic of spam and email that people have to pay to receive what are essentially advertisements. We don't approve of that in the electronic mail domain and I don't think it's a good idea with message bank or SMS [texting].'

However, as long as the commodity circuits of the two kinds of medium are kept separate, consumers are generally content to participate in both forms of commodification. In the context of the metro-nucleation wrought by the culture industry, it is easy for Internet Service Providers (ISP) to appeal to consumers concerning their interest in exploring an expanded range of horizontal communication mediums. For a time-charged fee, the ISP will electronically remove the cellular architecture which divides individuals from others locked into the same system of 'widely dispersed consumption points'. And of course the promise is that it will do so more comprehensively than a telephone company can and with much greater bandwidth. It is this feature which prompted Howard Rheingold (1994) to speculate as to whether the Internet would be the 'next technology commodity' (60–1). Certainly, an inspection of those dot.com stocks before the crash in the late 1990s would have had most of us being readily convinced by Mr Rheingold.

Understanding network communication in the context of broadcast communication

As has been argued, the distinction between first and second media age is a useful one to the extent that it suggests that 'broadcast' and 'interactivity' carry ontologically distinct forms of social tie, differences which are to a limited measure clarified by the epochal distinction. Sometimes, post-broadcast theorists glimpse the fact that this distinction need not be historical (see Baym, 2000; Wark, 1999). McKenzie Wark (1999) suggests, deferring to Innis as the classic exponent of this view, that '[t]elevision makes it possible to generate vast publics, attuned simultaneously to the same message; the telephone makes it possible to coordinate personal connections, exchanging particular and self-generated messages' (23). Wark sees these two technologies as providing a basis for different kinds of 'culture', which for him inhere in the messages rather than the mediums they allow: 'Through the television and telephone, quite different kinds of culture coalesce; one based on normative and majoritarian messages; the other at least potentially enabling the formation of marginal and minority culture' (23–4). In Wark's analysis, it is the dynamic between 'mainstream' and minority that is reproduced in these two technologies, a dualism that is in fact at the heart of second media age politics, from the visions of Apple users in the 1960s to the current utopian hopes of using the Net for peace and global protest against 'the system' (see Terranova, 2001).

Wark's discussion is a useful departure from succession models of media. He argues that there are at least two kinds of cultural productions arising from different media forms, and that these can co-exist in tension. He is clearly not arguing that television's 'mainstream' is progressively undermined by increased telephony or any other medium which makes it possible to coordinate personal connections, exchanging particular and self-generated messages.

What Wark avoids, in this discussion at least, is choosing technologies which are marked by considerable historical separation: TV versus Internet, cinema versus interactive television, etc. By appreciating a co-presence of communicative technologies and the social or cultural dynamics they produce, the historical distinction is weakened. Instead, it becomes necessary to appreciate the way in which 'interactive' and 'broadcast' forms of communication carry different kinds of social bond. These two kinds of social bond are mutually co-dependent as well as, in the current era, carried much more heavily by technological extension. It is this increased degree of extension which, it might be argued, could qualify as a second media age, much more than the predominance of interactivity over broadcast. If anything, technologically extended inter-activity has not eclipsed broadcast; it has merely provided everyday forms of interaction with increased alternatives.

As carriers of integration, broadcast and the Internet can broadly be described in terms of the predominance of mediated forms of either recognition or reciprocity. Both processes, collective recognition and extended reciprocity, carry with them modes of integration of persons which, as our discussion of mobile privatization suggests, prioritize the relationship of individuals to an 'outside' on the basis of commoditized social relationships.

In this, it can be argued, it is impossible to adequately understand the extended reciprocity of network communication without understanding the social dynamics of broadcast communication. As we have seen, mediums of broadcast integration provide the socio-spatial as well as the ideological preconditions of virtual communities. Broadcast communication, in establishing, as Adorno and Horkheimer show, a horizontal cellular architecture which can only be integrated via mass media, has heightened the dependence individuals have on those media, or on whatever horizontal means of communication that can break down the tele-mediated fields which divide each adjacent unit from the next.

In this connection, the notable decline in extended families and face-to-face neighbourhood kinds of networks corresponds inversely to the rise of 'metro-nucleation'. As Geoff Sharp points out, many years ago, in *Family and Social Network*, Elizabeth Bott (1971) clearly demonstrated the transition to a later modern mode of integration in which the reconstruction of the material habitat contributed to the increased segregation of the institutional settings of everyday life (Sharp, 1993: 236). To quote McLuhan and Fiore: 'The family circle has widened. The worldpool of information fathered by electric media – movies, Telstar, flight – far surpasses any possible influence mom and dad can now bring to bear. Character no longer is shaped by only two earnest, fumbling experts. Now all the world's a sage' (in McLuhan and Fiore, 1967: 14). Indeed the family circle has widened, but not before it has also shrunk – a contraction of embodied wide-kinship networks which then find substitution in the global openness of electronic media of all kinds.

However, whilst this dual operation of expansion and contraction wrought by broadcast media has increased the ability to identify with far greater numbers of other persons than was possible in pre-media societies, it has also dramatically reduced the daily physical interaction with such others. Such re-territorialization of social architectures of identification creates heightened, almost religious, attachment to extended media (see Martin-Barbero, 1997), which is inversely related to widespread breakdown of families as a basis for social capital in media societies (see Fukuyama, 1999). As face-to-face familial and 'local' associations with others become overpassed, in the manner of an inter-suburban freeway, interaction increasingly becomes confined to a level of association which has an abstractness which is somehow adequate to the technically mediated bond which abolished its local expressions – broadcast. As the embodied public sphere is replaced by an electronic public sphere, to 'go out', as McLuhan once remarked, is to be alone (cited in Levinson, 1999: 134). And when we are out, he noted, our interactions become more violent, insofar as we have been stripped of readily available roles. Road rage (Lupton, 1999), street rage, telephone rage, even air rage, become more common. Insofar as, McLuhan argues, the medium is our identity, when we turn off the Internet we lose the role it assigns to us as a netizen, and we are forced to take responsibility for our identity.

When we are immersed in the comfort of electronic media, on the other hand, we are able to foster new types of shared experience, resulting in 'greater personal involvement with those who would otherwise be strangers' (Meyrowitz, 1995: 58). However, the mobile privatization side of this connection, when we try to confront the disjunction between two kinds of interaction – face-to-face and disembodied – can also lead to Net flaming (see Dery, 1994; Springer, 2000).

In the modern era, the trends towards increased densities in, and increased use of, telephonic and Internet-based forms of communication are an outcome of the fact that the 'virtual community' of broadcast extends the sphere of recognition, but excludes interactivity as an ingredient of such a community. One of the most instructive signs of the inadequacy of broadcast in this regard can be found in recent changes to its genre. On television, it is no surprise that it is precisely since the availability of the Internet that 'reality' TV has became an established genre, still clinging to the last remnants of simulacra, the semblance of *undirected*, authentic, spontaneous and intimate personality, which other genres so comprehensively lack in the context of network. Changes to how we are able to control the viewing of television with digital video are also an example of new kinds of expectations about media brought about by the increased availability and normalization of interactive technology. A recent advertisement for DVD in Australia boasted the virtues of being able to view films as you have never been able to before. DVD 'puts you in the director's chair'. It allows you to 'get behind the scene' and pass through that wall which separates the producers and consumers of media, meaning, in some removed, imaginary sense, that we are able to 'interact' with the producers by being able to direct proceedings as they are.

But by far the most visible indication of the way in which individuals seek to overcome exclusion from interactivity in a public domain is in the extraordinary production of personal websites. Net enthusiasts of all kinds see the Net as a way of gaining visibility in a world where they are otherwise rendered as anonymous consumers. It provides an extended 'mediaplace' in the world which they don't otherwise have. A typical page conveys a level of intimacy via pictures and personal life which can't be achieved in institutional life, such as where you work or where you study. It seems appropriate to put your photo on the Web because others rarely appreciate you in the context you would prefer them to see you in. People from your working institutional life seldom appreciate who you are – the fact that you might be a good sportsperson, or a fine cook. To the extent that personal web-page authors believe they have a visibility comparable with broadcast on the WWW, they can deliver themselves from a feeling of anonymity.

However, the *experience* of visibility is inevitably limited by the very architecture of Internet communication. As Schultz (2000) points out:

> Bulletin boards and Internet discussion groups can balance the power and biases of traditional mass media and play an important role in controlling and criticizing journalism as well as in establishing mobilizing types of communication … . As Friedland … has suggested, the Internet gives people

a fine tool for an 'electronic public journalism' that is independent from professional media organizations. However, these new opportunities involve problems that former media critics really did not face. How attractive is the content of such a Web page compared to the highly produced content of broadcast? Communication and participation alone do not mean much in terms of quality and value of content. Also, communication can remain without any significant effects as long as it is not transformed into communicative power and effective decisions ... . Eventually, there is a seemingly trivial but most important consideration: the greater the number of communicators, the less time everyone has to listen to others; the smaller the size of interacting groups, the smaller their significance for society as a whole. This is one of the reasons why one must doubt whether Internet enthusiasts are right in their belief that the end of traditional institutions of politics and media has come. They suggest that a new elite of 'netizens' is going to take over society ... . But on what integrational foundations is the alleged net community grounded? There seem to be few apart from an individualistic rhetoric of free information and a euphoria about thousands of subcommunities to which no one can belong at the same time anyway, not even in bodiless cyberspace. After all, attention is one of the most valuable resources in the new era ... . Economists would call it a very scarce commodity. With a growing number of information and communication forums, some central sources may become more important. They can reduce complexity, help users make judgements about what is important, and build shared beliefs. (207)

Understanding broadcast communication in the context of network communication

By far the greatest contribution of the second media age thesis is that it acts as a powerful lens for analysts of media to understand something about broadcast which has, up until now, been difficult to see – its character as a medium. The mere fact that 'television' as the standout techno-cultural form of broadcasting has recently become *formalized* as a distinct domain of analysis is significant in this regard (see Casey, 2002; Corner and Harvey, 1996; Geraghty and Lusted, 1998; McNeill, 1996; Newcomb, 2000).[10] Television studies, as a sub-discipline of media studies, has acquired a new positivity. Television has come to mirror the way in which the Internet either has become a distinct technology of communication, or is posited as a stand-in for broadcast-in-general, just as the Internet emblematizes the rise of the network society.

Throughout most of the period in which media studies thrived in its analysis of broadcast, by looking at ownership and control, media institutions, media content (from semiotics to ideology and hegemony) and, latterly, audience studies, the one area that was left the most neglected was that of broadcast as medium. With the exception of a small burst of medium theory in the 1960s and 1970s, linguistic and semiotically based accounts of media dominated all of the different schools, from conservative 'effects' analysis to radical Marxist accounts of ideology.

The value of the second media age thesis is that it makes the study of broadcast *as a medium* all the more distinct. On the other hand, what is obfuscated by the second media age argument is the fact that this distinction cannot be so easily periodized. But, inevitably, attempts are made to do this by claiming that New Media are signalling the demise of old media, or at least changing the way in which the old media are related to. Indeed television and the Internet are most commonly posited as offering the sharpest contrast by which this periodization can be established. In this scenario, all media are reduced to two standout forms. Either the Internet is seen to overtake television, or it has changed the culture of television. In the latter case, for example, Bruce Owen, in *The Internet Challenge to Television* (1999), argues that television is beginning to change at the same pace as computer networks. Television viewers, he claims, now accept the far higher level of freneticness which is commonplace in nearly every rapid-cycle television advertisement we watch.

But some have attempted such temporalization without taking on a second media age position. In his essay 'What Was Broadcasting?', David Marc (2000) argues that, in the US context, '[t]he Broadcast Era, a period roughly stretching from the establishment of network radio in the 1920s to the achievement of 50 percent cable penetration in the 1980s, becomes more historically distinct every time another half dozen channels are added to the cable mix' (631). Marc's argument is that with the introduction of cable television in the USA, the niching of broadcast leads to its de-massification, and the collapse of its transdemographic possibilities, which will consign it to a 'biblical era of mass communications' in which great events and famous people will have become entombed (631).

Marc's argument, based solely on the US experience, that 'mass broadcast' constitutes large publics and pluralized forms of broadcast constitute multiple public spheres is significant from the point of view of reinforcing the argument that broadcast is a constitutive medium. Regardless of how large or 'transdemographic' it is, broadcast constitutes audiences within particular fields.[11] The changes in television broadcasting by the mid-1980s in the USA were indeed significant to the reconfiguration and distribution of audiences, but did not in themselves cause any major rethinking of what broadcast, as a communicative form, actually is.

What is significant in the period of the inception of the Internet, therefore, is the sudden return of interest in medium theory as a legitimate perspective in media studies (see Chapter 2). From the early 1990s onwards, when the Internet began its exponential growth, the theoretical necessity of analysing the social implications of communication mediums had become paramount, if not unavoidable.

A key reason why medium theory has rebounded is to do with the fact that the Internet and CMC are so much more visibly seen to be a social medium than is broadcast. From a McLuhanist perspective, this is paradoxical insofar as he argues that new medium environments remain unperceived 'during the period of their innovation' (McLuhan and Fiore, 2001: 17). However, it is precisely because, in an everyday sense, the Internet is seen as a tool, or as a vessel/conduit 'highway' (see the discussion below of Meyrowitz's three metaphors of media), rather than an environment, that it is seen as a medium much more so than is broadcast. An appreciation of how the Internet might be a medium-as-environment is less common in an everyday sense. Most studies of the Internet examine how it is a means of connection, a superhighway for virtual travel, or a mode of association that makes possible virtual community defined through *connecting individuals who have similar interests.*

In terms of the foregoing discussion of the different qualities of technological extension that are manifest in different communication mediums, it is useful to outline some of the qualities of broadcast which are not possible on the Internet – all of which have to do with the *communication event.*

Broadcasting and 'datacasting'

We have already suggested that broadcast needs to be considered as a form of socio-communicative bond rather than a technical medium. In this way, it is necessary to appreciate the fact that broadcast may or may not be technologically extended. However, across the technologically extended spectrum of broadcast, it is possible to distinguish different types of broadcast event according to its visibility and the synchronicity of its audience. In doing so, we can see how datacasting via digital broadcasting services and on the Internet is not quite the same as conventional broadcast.

- Datacasting is a service that has long been offered by digital television providers, but in this context its range of qualities is more variable: multiple kinds of audiences can be constituted within such a transmission platform. However, the fact that the 'live' forms of the transmission have subscribers, unlike 'datacasting' on the Internet, means that the digital television version of datacasting retains a 'mass constitutive' function.[12]
- There are also more recent forms of 'interactive' datacasting which enable the streaming and caching of video and audio services that have found their most attractive application in 'video-on-demand'. These services are 'invocational': the consumer chooses the time to call up the content, but not as part of an audience.

- It is also possible to use broadband Internet for interactive datacasting of televisual content, but as its only proven value is in delivering video-on-demand, this value is limited by much smaller screens.
- Other categories of Internet datacast, such as a simple broadcast email, webcasting or bulletin board posting, are asynchronous as they are mediated by the storage of media which are retrieved at indeterminate and undirected times by the eventual audience. Nor do social expectations about Internet datacasting conform to any kind of appointed regulation, unlike broadcast events.

Only when an Internet datacast is parasitic on already existing broadcast events does it have this quality. For example, when the Olympics are being beamed around the world, mirror-sites on the World Wide Web will assume a high visibility also. However, where Internet datacasts are not paralleling media events, they will only ever have an irregular visibility. It is only when a datacast is itself reported in a broadcast that it will acquire a visibility comparable to the form of media in which it becomes known.[13]

For example, as remarked above, it is impossible to become famous on the Internet in any of its sub-media. The only persons who have become famous were ones reported by the mass media as being popular Net personalities. In truth, however, the reason why their web-pages have received such attention is precisely the exposure they have had in broadcast. Fame on the Internet can only be 'second-order' fame which parallels an audience that has already been constituted by a synchronous, highly visible media event of some kind.

Gauntlett (2000) draws on an argument from Michael Goldhaber that visibility on the Internet is reduced to an 'attention economy'. *Attention* is a scarce resource (Gauntlett, 2000: 9). If a website does not have 'interesting content', other websites will not link to it. However, Gauntlett fails to explore what determines a site's content as 'interesting'. He correctly points out that the amount of money backing a site is irrelevant to its popularity, but ignores the power of broadcast media in making this determination. Thus he gives examples of 'penniless', 'ordinary' people who are supposed to have accumulated large amounts of attention on the Internet:

> To take a real-life example, Harry Knowles – an ordinary, hairy, twenty-something guy from Austin, Texas – has received much attention with his *Aint it Cool News* (www.aint-it-cool-news.com), a website providing daily Hollywood gossip and movie previews from a network of 'spies' (industry insiders and people who infiltrate test screenings). ... Knowles is now very well known and much in demand. (11)

There are numerous layers of Gauntlett's discourse that can be unpacked here to show that Knowles' fame has nothing to do with the medium of the Internet, but everything to do with broadcast.

Table 4.1 The broadcast event

	Regular-visibility broadcast	Irregular-visibility broadcast
Synchronous audience	Public speech Television Radio Daily newspaper	Media stunt Newsflash Baudrillard's 'obscene' (see p. 107)
Asynchronous audience	Magazine Novel Billboards	Datacasting Computer virus

- The characterization of Knowles as 'ordinary' already reaffirms a basic binary upon which the symbolic inequality of broadcast operates, between high and low visibility.
- It is a real-life example supposedly because it did not occur via 'the image', but in fact Knowles' attention was achieved entirely by trading in images, and the system of images, for which he became an agent.
- Knowles' 'fame' is entirely parasitic on the Hollywood personality system, which is the basis of his site being 'interesting'. It is not because Knowles is hairy, twenty-something or parodying another dominant genre – news.
- In turn, Gauntlett publishes Knowles' URL, which may receive more visits as a result of people reading Gauntlett's mass-produced book.

Moreover, a website's popularity will vary over time depending on its synchronicity with events managed by the mass media, and made visible by mass media, for which it can only ever be a mirror, or an anti-site (a site which reaffirms the power of a broadcast text in its efforts to parody or criticize it).

Table 4.1 classifies forms of broadcast by their synchronicity and visibility. A broadcast event may be asynchronous but have high visibility, such as magazines and billboards. They tend to have a regular but not very immediate visibility. The impact of the content of such forms tends to build up over time. A broadcast event might also be synchronous but have irregular visibility, such as a media stunt. Such stunts, sometimes attempted by social movements, tend to be highly irregular, but, just for that reason, acquire high visibility and a synchronous audience when they do appear. Some broadcast 'mega-events' exhibit a dramatic concentration of both visibility and synchronicity (see Garofalo, 1991; Real, 1984). No event collapses more of the positive features described above than do the Olympic Games – the most 'widely shared regular event in human history' (Real, 1984: 222).

Whilst there are these variations in kinds of broadcast media, the way broadcast differs from interactive media is far more significant in understanding contemporary media. By such a contrast, a host of qualities of broadcast media come into view which were previously difficult to see.

Broadcast is the only medium that is capable of being 'live'

> The medium alone makes the event – and does this whatever the contents, whether conformist or subversive. Serious problem for all counter-information, pirate radios, anti-media, etc. (Baudrillard, 1983: 101)

In the section above on technological extension we discussed how, in enabling the continuation of face-to-face kinds of cognitive communication across time and space, electronic and broadcast media also introduce entirely new qualities of interaction which are not possible within face-to-face communication.

Datacasting on the Internet lacks one paramount quality that television and radio and newspapers possess – that of being 'live'. When we say an image is 'live', it usually refers to the fact of there being no separation between the time of production of a message and the time of its reception, such as the coverage of a sporting fixture. This is known as 'real-time' and may exist in broadcast or interactive contexts. However, there is a sense in which only electronic broadcasts are always 'live'. Whatever the temporal origin of the content of a media broadcast, what makes it live is the fact that it is simultaneously being experienced by a mass audience whose very quality *as* a mass is constituted by the broadcast itself.[14] The intensity of the 'liveness' operating in a broadcast event is also related to the size of the audience, and its capability of immediacy – the fact that it may be showing real-time footage or that such footage might be able to intercede at any moment. Much has been made of the difference between cinema and television precisely around the liveness differential. Flitterman-Lewis (1992) suggests that television succeeds in the production of 'presentness' in a way that film cannot. It offers a 'here and now' in opposition to cinemas's 'there and then' (218). Television's 'peculiar form of presentness' founds its triumphal claims to immediacy:

> You should think of television performing its most distinctive function, the live transmission of events. ... Unlike cinema the sequence of the actual event cannot be reversed when shown on television. ... The now of the television event is equal to the now of the actual event in terms of objective time, that is, the instantaneous perception by the observer of the actual event and by the television viewer. (Zettl, 1973: 263)

Zettl's specification of this distinctive function, however, goes only part of the way towards understanding *broadcast*. In casting a television broadcast in terms of a representation of a live event, questions of realism, and bandwidth (i.e. TV versus radio) also necessarily come up. The point is successfully made that only in such a transmission do the time-world of the represented event and the time-world of the viewership coincide. However, the 'nowness' view of 'liveness' falls short of understanding the role of the audience in media events, and the fact that television programming also provides a context for 'liveness'. In her essay 'TV Time and Catastrophe, or Beyond the Pleasure Principle of Television', Patricia Mellencamp (1991) argues that the distinctive quality of TV time is not that it is capable of simultaneity with live events, but that, in a Baudrillardian sense, it is simultaneous with itself, and other programmes that happen at the same time. For Mellencamp, when coverage of a catastrophe interrupts the regulated half-hour viewing of daily transmission, the thrill of interruption produces the very liveness of the event. Instead, the regularity of the scene is menaced by what Baudrillard calls the obscene, which is the co-product of the graphic, the sensational, which shocks audiences into the hyperreal.[15]

This idea is also found in Dayan and Katz's analysis of media events as 'high holidays' away from the routine of programming. These events have a special place themselves in the history of broadcast media in that their importance has coincided with the globalization of media. What once captured the attention of a nationwide audience can quickly progress to worldwide status. But for Dayan and Katz (1992), this need not be the televising of catastrophe, but such events typically include contests, conquests and coronations – 'epic contests of politics and sports, charismatic missions, and the rites of passage of the great' (401).

What is distinctive in Mellencamp's analysis is not the relationship between a media message and an individual viewer, listener or reader, but the fact that each member of a media audience is aware of the reach and cross-contextuality of the broadcast, and as such the media event takes on a power beyond the meaning of the individual messages.[16] Marshall McLuhan was perhaps one of the most powerful exponents of this quality of broadcast, which, in his case, was a characteristic of what he viewed to be auditory culture. It is when information becomes instantaneous and comes from all directions that it impacts in tribalizing ways. For this reason the assemblages and information that are possible within even visual electronic media and genres like newspapers are in fact 'aural', in McLuhan's terms. Newspapers, which graphically arrange information in a non-linear fashion, are in fact, in McLuhan's sensorial ontology, based on the medium of the ear and its sensitivity to media which surround their audiences with a presence and range of extension not equalled by visual media (see McLuhan, 1964).[17]

The duration of a broadcast itself constitutes an event that is quite independent of the fact that its content may not have been produced at the same time. Electronic broadcast immediately qualifies for this effect, whereas newspapers qualify in a more limited sense. For the day of a newspaper's production it remains an event; indeed, Hegel once described it as the 'morning prayer' of modernity: 'The newspaper, Benedict Anderson says, is a "one day bestseller". Nobody reads last week's newspaper, unless they find it wrapped round potatoes in the kitchen. But everyday it sells out in millions' (Inglis, 1993: 29).

The predictions in the early 1990s that, as a result of making newspapers available on-line, print versions would become redundant within ten years have not materialized. For the same reason, the capacities of digital television (themed content, pay-per-view, video-on-demand) will never usurp the sociological appeal of broadcast, despite claims that this can happen simply through regulatory practices (see Steemers, 2000). Newpapers, like television and radio, make possible an aspect of communicative solidarity which Net mediums cannot, and will never be able to fulfil – the fact that they are able to be performative.

Broadcast is the only medium that can be performative

The performative quality of broadcast is an extension of *one* quality of speech, which the language philosopher J.L. Austin labelled the 'speech act'. In everyday speech, an utterance may be considered an 'act' when it refers to an immediate situation. The utterance does not have to be 'true' or 'false'; rather, it becomes a form of action which has mutual consequences in a setting of 'live' interaction. In speech act theory this situation is usually part of a face-to-face *interaction*, where utterances refer to the present-at-hand in the form of 'here' and 'this'. Anyone outside the range of a speech act will not be able to interpret its meaning. Conversely, those within the range of the speech act will potentially feel part of an exclusive *speech community*. Such a speech community of mutual presence will be able to realize its distinctive group dynamic the more speech acts are made. Actions such as promising ('we will do that later on'), naming ('this is the best …'), warning ('watch out for …'), requesting, insulting, and greeting have a different meaning in relations of mutual presence than, say, in writing or in Internet interaction.

The important thing to stress here is the degree of mutuality, or how many other people are *simultaneously being acknowledged* as hearing a speech act. When speech acts are formalized into *speech events*, like lectures, public talks or indeed speeches at formal gatherings, the boundaries are also formally defined and generally known. In such circumstances, the extent to which the audience will know something of the speaker will itself add to the meaning of the speech. However, whilst there are occasions when little may be known of the speaker, one characteristic is common to all speech events – the fact that a given speech act is constitutive, regardless of the content, of an audience. It is the constitutive function of extended speech acts across time and space which makes possible public opinion also. Public opinion is entirely an outcome of the performativity of communicative fields. Such opinion does not issue from the mass, except by and for the institution of radial communication. Public opinion is merely a reflex of the mobilized or formulated forms of organized discourse endemic to the structure of a performative apparatus. Outside of this, public opinion, as Pierre Bourdieu (1993) once famously suggested, 'does not exist'.

This latter quality is central to understanding broadcast. The performativity of broadcast derives not merely from its 'live' quality but also from the fact that it can technically extend speech events to an almost unlimited audience. When the news presenter says that 'the world is in shock' because of the death of a member of the British royal family, there is a sense in which this utterance *is* the event, more so than the actual death. Of course, the utterance may rest on the truth of a state of affairs, but what is never in dispute is that, at a certain hour, such a statement was made to the world. When we are told the world is in shock, we, as audience members, are immediately enveloped by such an utterance, regardless of our attachment to the deceased royal. We may as well *be* in shock in the sense of consummating what is likely to be the common state of all audience members. The outpouring of mourning for Princess Diana was almost entirely an effect of the powerful performativity of media.[18]

Similarly, recall from Chapter 2 the widespread panic over a Martian invasion as a result of Orson Welles' radio broadcast of *The War of the Worlds*. As a novel, H.G. Wells' book could never have the same effect precisely because it lacked the synchronous audience which electronic broadcast and newspapers provide. Similarly, extended-interactive communication can never constitute an *event* as broadcast does. Indeed broadcast can *be* the event. This is particularly salient with broadcasted mega-news, which takes on an historical status far more powerful than any pre-broadcast events would ever allow.

> The audio-visually documented assassination of John F. Kennedy levitated him and his presidency into an historiographical mythosphere once occupied only by the likes of Washington, Jefferson, and Lincoln. By contrast, William McKinley never had a chance. The O.J. Simpson trial is assured a place in the popular history of the twentieth-century American jurisprudence that few, if any, Supreme Court rulings might hope to occupy; Judge Ito has already eclipsed Oilver Wendell Holmes in recognition factor. Father Coughlin will have generated far more usable material than Pope John XXIII, and Billy Graham more than both. Who is likely to be a more dominant presence in the digital archives? Albert Einstein or Carl Sagan? Dr Freud or Dr Ruth? Charles Darwin or Pat Robertson? Mother Teresa already has a higher F-Score than Albert Schweitzer. In the future the past will belong to the audio-visually reproducible. The giants of the arts and sciences who, for whatever reason, failed to climb the transmission towers of the twentieth century can expect to be remaindered to the specialists' bin. (Marc, 2000: 630)

Marc argues that it was the culture of broadcasting that made Elvis and the Beatles possible: 'They emerged from the night spots of Memphis and Hamburg to appear on *The Ed Sullivan Show*. In the cable environment, however, *The Ed Sullivan Show* is no longer possible' (630).

The foregoing examples serve to show that broadcast and network communication are ontologically distinct – a distinction which has numerous consequences for the kind of telecommunities that technologically extended communication makes available. This is a topic which we shall return to in Chapter 6.

Only broadcast is constitutive of a media 'mass'

Another difference between the first and second media age, or, as this book proffers, broadcast and network architectures, is that whilst both produce individuation, only broadcast enables a mass *conscience collective*. As has been argued elsewhere, while the Internet facilitates reciprocity with little or no recognition, broadcast facilitates mass recognition (in the form of meaning integration) with little reciprocity. Moreover, the mass is constituted as an effect of the broadcast apparatus itself.[19] There is no mass without broadcast, in the sense that it is a social body entirely derived from extended speech events. Such a mass has little opportunity for horizontal interaction between each of its members, nor is there much opportunity for interaction between a mass and the few centres of media production, but the mass itself retains its basic form and solidarity, Althusser has argued, as long as it accepts its interpellation via such a centre (see Chapter 2). Broadcast so homogenizes a centre and makes possible the so-called 'mainstream', that it creates within itself a tension between the impulse for collective totemization of an image, a message, a narrative (the synthesization of a *conscience collective*), at the same time as it reinforces the cult of individualism from which the need for such synthesis is drawn. Broadcast effects a dual movement of separating and uniting a given audience, which divides the reciprocal cohesion of subjects (based on direct interaction) and reconstitutes this cohesion on a more abstract basis of association.

As per Table 4.1, the solidarity of a media mass depends on the kind of broadcast. The more performative the broadcast, the more cohesive is the *conscience collective* of the mass formed by such an event. It is also evident that this is not simply an effect of such broadcast but is itself 'popular'.[20] It is the performative feature of broadcast which can give it a bardic function, as discussed by Fiske and Hartley (1978) in relation to television.[21] Individuals wanting to find out about a spectacle event that they have heard about (from a friend or a fragment of news) will inevitably tune in to a big network channel. From the standpoint of the broadcasters, Marc (2000) also points out the sermonic features of modern broadcast: 'Broadcasting by its nature is an evangelical activity, whether it is used to preach the gospel of consumerism (commercial television) or the gospel of ethnical culture (PBS [Public Broadcasting Service]). A broadcast typically invites everyone who can receive its messages to sympathize, empathize, learn the creed, buy the products, and join the fold' (640).

However, Marc, who adopts a technologically centred understanding of broadcast, views cable television as not capable of such bardic functions, and suggests that the decline of over-the-air transmission can be blamed for the 'severe social dislocations and polarizations that have taken place in American society' (640). The 'nation-as-audience' is seen to be overcome by a 'nation-of-audiences'. It is true that cable television, or narrowcasting, does not, by definition, have such large mass audiences, but they are just as capable of constituting a mass, just as free-to-air television is. This can happen over time, or by the fact that narrowcasting content is seldom produced just for a particular channel, but is generally reproduced and reproducible on potentially global scales, in duplicate narrowcasts or broader broadcasts. 'Indiscrimination', Marc (2000) insists, is the characteristic which distinguishes broadcast from cable (640). However, as more recent paradigms in audience studies demonstrate, discrimination is the preserve not of the broadcaster, but of the audience. Whether audiences discriminate between *spatial* locations on the cable dial, or between *temporal* slots in broadcasting programming, is not nearly as significant as the fact that, in each case, a mass is constituted by radial transmission. Structurally, whatever the mix of broad versus narrow transmission, the nation-of-audiences will always prevail, and whatever the mix, the nation-as-audience will always be possible should a spectacle arise that is presupposed by such a form. A day in September 2001 made this plain ... and not just for a nation.

Broadcasts that have regular and stable visibility can become points of ritual attachment, be this a weekly soap opera, the nightly news, or a once-a-year fix on a sporting tournament. Where the performativity of a broadcast event is at its highest intensity, it is even possible for the media-generated fields which atomize audiences at a face-to-face level to momentarily disappear.

When Princess Diana's death was announced in 1997, numerous reports were made of persons who never even followed the life of the Princess but were overcome by spontaneous grief, whatever they were doing and wherever they were. At the moment the news was received of the death of the most photographed person in the world, the local contexts of association which would otherwise occupy them were rapidly dissolved, as the most binding field of recognition became the calling of global media.

However, typically, the synthesis produced by broadcast media rarely produces such affective horizontal attachments, as audiences need only look to ongoing media narratives to consummate a sense of belonging to a telecommunity.

Audiences without texts

There is no audience outside of broadcast. The performativity of a broadcast event – whether it is synchronous or asynchronous – and its instantaneity are much more important factors in the formation of an audience than the kind of text which it communicates. Genres and texts may be organized in repetitious ways that facilitate strong identification with a common language of meaning, but a large part of an audience will be drawn to a broadcast event simply out of its performativity. In fact, it is possible for some audiences to form *only* out of the performativity of a broadcast. In such cases, the idea that everyone is consuming the same message is more important to audience members than are the affect and meaning of the text. This is especially true if the media event is synchronous. This adds a dimension of simultaneity to media consumption, or, as Hills (2001) argues, in applying Benedict Anderson's (1983) concept of 'imagined community' to fan communities: 'The kind of imagining characteristic of a nation is therefore one of pseudo-simultaneity, the assumption that thousands of anonymous, unseen and unknown individuals are watching the same television programme at the same time' (Hills, 2001: 152). This simultaneity effect is particularly true of spectacle events and breaking news, but it is also true of channel surfers, who are immersing themselves in the medium much more than in any narrative content.

The fact that the formation of a given audience rests, in whole or in part, on the performativity which a broadcast medium supplies to the texts which are communicated in them problematizes developments in audience studies which advance the oxymoronic notion of the 'active audience' (see Nightingale, 1996: 7–8). In this assignation audiences are active users of mass media insofar as broadcast messages are useful. Paradoxically, the postmodern thesis on the active audience rests largely on the behaviourist premises of the uses and gratifications approach to audience research (see Katz et al., 1974).

The account of the active audience is ultimately tautological insofar as audiences do not antedate broadcast events, but are internal to them. The active audience argument conflates the audience of a particular technical medium like radio or print with patronage of the texts and genres of these media. To say, for example, that radio audiences are 'diversified' is to misconceive an audience as abstractly belonging to a technical medium, rather than to a particular media event (which may happen to be made possible by radio, television or print). This can be seen from the realities of actual audience measurement. Few audience survey instruments are ever conducted in relation to a technical medium, only in relation to actual media events, such as an edition of a newspaper or magazine, or the screening of an electronic media programme.

In UK audience research, such as Dave Morley's study of the UK current affairs programme 'Nationwide' (1980), the active audience was defined in relation to interaction with texts rather than mediums. Morley's analysis drew on Hall's encoding/decoding schema, which allowed for negotiated and oppositional readings, rather than just the usual dominant reading that in implied in processual models of communication. Virginia Nightingale (1996: 16) suggests that Morley's 'audience' might best be described as a constituency – a group defined by its common use of the same signs and codes (be this critical, negotiated or passive). In this case, for Morley, the audience, as Nightingale explains, has a kind of 'dual existence' – 'it is part of the mass audience and also part either of subcultural or communal relations with others' (Nightingale, 1996: 15). Interestingly, Morley turned away from the audience–text interaction research of 'Nationwide', to look again at audience–medium interaction in later work (see Morley, 1992).

Paralleling such a shift was the attempt to locate the audience. Is the audience the same as 'the mass', or can it only be found in texts? The mass argument is criticized by theorists like Ang (1996) who argue that the audience-as-object is an invention of media institutions and corporations. The entire superstructure of marketing exercises that audiences need to be 'reached' is founded on this myth. Audiences are not ready-formed receptacles awaiting to be discovered but are constituted in the same operation as the audience–text interaction.

Another sense in which audiences are possible without texts is related to the way the growth of network communication has been accompanied by heightened levels of 'audience participation'. This is especially evident in the spectacular rise of 'reality TV' programmes and spectacle features – these are examples of unscripted content (see Couldry, 2003: Chapter 6). Of course they are still social texts in the formulaic way in which they unfold, but they work much more on the principle of imaginary substitution – the suggestion to audience members that they are seeing (a representative of) themselves on the other side of the screen, or at least can imagine themselves doing whatever is being screened. Once again we can see how a new genre allows us to reflect on old mediums that have become acquainted with their ontological power.

The return of medium theory

> Man, [McLuhan] understood the internet in the sixties. He *was* the internet in the sixties. The world's just finally caught up to him. He was the internet in the sense he was in touch with the entire globe. ... He was wired long before the editors of *Wired* magazine were born. This man was truly wired. (Robert Logan in Benedetti and Dehard, 1997: 171)

In the previous chapter, we discussed the return of McLuhan, who, in being the most prominent first-generation medium theorist of electronic broadcast, has enigmatically become the subject of his own prediction – that theoretically we can view old mediums from the vantage point of the new, but that at the same time we attach ourselves to new media through the objects of old media – what he called rear-view mirrorism.

In this chapter, I have suggested that both network and broadcast can be viewed from the standpoint of each other. This is a corollary of the fact that I have sought to reject a linear succession model of media forms. In doing so, broadcast and network communication can be looked at from the perspective of medium, and each of them can become the basis for revealing the limitations of the content views of both forms of media.

One of the obstacles to fruitfully comparing and contrasting the two kinds of social medium that are carried by network and broadcast is the fact that the discourses around 'medium' are, typically, technologically reductive. Certainly, it is commonplace to think of the Internet as a medium in a way which is contrasted with television as a medium.[22] Similarly, distinguishing between technologies within the same sociological medium has the same effect (for example, see Marc's [2000] discussion of radio and television as separate mediums).[23] To contrast television with the Internet sharpens a period distinction between media forms, but softens and obscures the analysis of broadcast and network as distinctive integrative 'shapes', as discussed above. However, this relationship is seldom adequately theorized. For example, even though McLuhan is heralded as having 'understood the Internet' in the 1960s, the Internet in itself is not a medium in the sense in which medium is explored by McLuhan. For McLuhan, as we have seen in the previous chapter, mediums are much more tied to how they extend individual bodily senses. Some technologies extend just one sense (telephone – the ear; print – the eye) while others extend numerous senses at once (cinema, television and the computer, in which a variety of hot and cool 'effects' are produced). Moreover, McLuhan's paradigm has the (perhaps accidental) virtue of grouping broadcast technologies according to the personal-social experiences they produce. Thus TV is considered to be an acoustic medium like radio, in which sound represents the privately experienced equivalent of a social world characterized by 'information from all directions'.

However, the Internet as a form of 'cyberspace' increasingly comes to be termed a 'medium' in itself. For example, Christopher Horrocks (2001) argues that McLuhan's insights 'now seem appropriate for a medium that arrived just after his message faded' (5). Although Horrocks' essay 'explores the encounter of Marshall McLuhan's major insights into media and technology with the present world of information networks, e-commerce, digital technology and the age of virtual reality' (4), the project of defining the Internet as a medium is left unaccomplished.[24] If anything, Horrocks is able to stress the continuities between broadcast and interactive communication through his exploration of the meaning of virtuality. For McLuhan, interactivity does not have to be defined as interaction between individuals through a medium, but can be defined as interaction with a medium itself. We might recall McLuhan's discussion of hot mediums as rich in information intensity (approximating virtual reality) but low in interactivity. Cool mediums require higher scales of interactivity from listeners/viewers/users. Thus, to follow McLuhan's definition, interactivity is common to both the first and second media age, but contention remains as to whether, with New Media, individuals desire to simulate a face-to-face environment or are content to interact only with the medium itself.

In Arthur Kroker's view, interactivity is not important; rather, what is significant is that we live in a processed world in which all individuals are essentially x-rayed by media:

> For McLuhan, it's a processed world now. As we enter the electronic age with its instantaneous and global movement of information, we are the first human beings to live completely within the mediated environment of the technostructure. The 'content' of the technostructure is largely irrelevant (the 'content' of a new technology is always the technique which has just been superseded: movies are the content of television; novels are the content of movies) or, in fact, a red herring distracting our attention from the essential secret of technology as the medium, or environment, within which human experience is programmed. It was McLuhan's special genius to grasp at once that the content (metonymy) of new technologies serves as a 'screen', obscuring from view the disenchanted locus of the technological experience in its purely 'formal' or 'spatial' properties. McLuhan wished to escape the 'flat earth approach' to technology, to invent a 'new metaphor' by which we might 'restructure our thoughts and feelings' about the subliminal, imperceptible environments of media effects. (Kroker, 2001: 56–7)

But what of the distinction between broadcast and interactive solidarity? If broadcast is also a medium of interactivity, what is the standing of these two forms as 'mediums'. From McLuhan the *Wired* magazine editors have taken up the idea of the Internet as an extension of consciousness itself. Horrocks (2001) quotes McLuhan: 'The next medium, whatever it is – it may be the extension of consciousness – will include television as its content, not as its environment ...' (pp. 52–53). Here McLuhan suggests television is itself a medium, and that whatever supersedes it will interiorize it. Certainly, cyber-utopians celebrate the idea that the World Wide Web is a place where every netizen can broadcast their own moving video or digital images. Paul Levinson (1999) suggests that the Internet is a 'meta-medium' which includes 'the written word in forms ranging from love letters to newspapers, plus telephone, radio ("RealAudio" on the Web), and moving images with sound which can be considered a version of television' (37–8).

This problem of medium is fruitfully explored by Joshua Meyrowitz in his essay 'Understandings of Media' (1999). Meyrowitz argues that three key metaphors have prevailed in the thinking of medium: medium-as-vessel/conduit, medium-as-language and medium-as-environment (44).

The first kind of metaphor, medium-as-vessel/conduit, is the most common. It is a metaphor in which a medium is regarded as a container for sending or storing content. It leads people to ask: 'What is the content? How did the content get there? How accurately does the media content reflect "reality"? How do people interpret the content? What effects does the content have?' (45). For Meyrowitz, this metaphor is so prevalent because it appears to transcend both mediated and unmediated interaction. It provides for intentionality across different media: 'We all have a sense that a message that someone loves us has power and meaning apart from whether we receive it in face-to-face interaction, by letter, by phone, by e-mail, or by videotape' (45). The message seems to be much more important than differences between the media. Using this metaphor, people believe that a movie can be made of a book, or that an interview can be transcribed into a journal article which has the same meaning. When we relate a television programme to a friend who hasn't seen it, we recount the content, the plot, the actors, etc. Meyrowitz points out that the greater part of media studies have adopted this metaphor with its focus on content and thereby overlooked medium directly.

The next two kinds of metaphors *are only found in mediated interaction.* Medium as language treats a medium *like a language* with its own grammar. This is a grammar of production variables such as font type, camera angle, sound reverberation, which are peculiar to mediated interaction, or, as Meyrowitz proffers, 'it is impossible for us to "cut to a close-up" or "dissolve to the beach" in everyday interactions' (47). Conversely, we consciously associate the meaning of a word, image or sound with its grammar or presentation. In a war movie, for example, 'we rarely see prolonged close-ups of "the enemy" but see numerous of "our soldiers", who, regardless of their actions, we sympathize with, as does the lingering camera' (48).

The last metaphoric perspective which Meyrowitz discusses is of medium-as-environment.[25] This is different from a media 'container' or conduit which carries or transmits a message. Rather a medium-as-environment comprises the fixed characteristics peculiar to a medium that make it unique, 'regardless of content and grammar choices' (48). These relate to (1) the type of sensory information the medium can transmit, (2) the speed and immediacy that is allowed a communicative event, (3) whether it is uni-directional, bi-directional or multi-directional, (4) whether the interaction is sequential or simultaneous, (5) the physical requirements for using the medium, (6) the ease of learning to use the medium.[26]

Like McLuhan, Meyrowitz maintains that individuals are generally not aware of extended mediums and the way they shape experience and perception. It is possible to be conscious of medium grammars, and when they are pointed out, they can be seen easily. Medium-as-environment is a different matter, however. Meyrowitz would agree with McLuhan that '[e]nvironments are not passive wrappings, but active processes which work us over completely, massaging the ratio of the senses and imposing their silent assumptions. But environments are invisible. Their ground-rules, pervasive structure, and overall patterns elude easy perception' (McLuhan, 1967: 68). However, unlike McLuhan, Meyrowitz stresses that to understand an individual communicative event, all three kinds of medium metaphors need to be considered together, but that they are typically considered in isolation, by distinct research communities.

Elsewhere, Meyrowitz suggests that the second and third kinds of metaphor began to inform a whole second generation of medium theorists in the 1990s including their thoughts on cyberculture. In 1994 Meyrowitz suggested that 'second generation medium theory' began in the late 1980s with a diverse range of analyses of media technologies. He pointed to the fact that a diverse range of communication forms was beginning to be looked at afresh from medium standpoints. Re-appraisals of photography, computer networks, 'smart machines' electronic text, TV versus print, and the relationship between communication technology, public space and social change are listed as fields of second wave medium studies (Meyrowitz, 1994: 69).

This second wave has consolidated itself in the renewed interest in the work of McLuhan, Innis, Carey, Katz and others. Perspectives explored in the previous chapter – on virtual community, virtual space and the renovations of CMC analysis – can be counted as part of this wave. More recently, the writings of Michel de Certeau, Margaret Morse and Karen Knorr-Certina suggest new and important directions in medium theory. But the explosion of literature on the interrelationship between New Media and the sociology of time and space is also central here. Concepts of time-space compression and 'technological space-time' (Harvey, 1989; Virilio, 1997), including 'cyberspace time' (Lee and Liebenau, 2000; Nguyen and Alexander, 1996), are part of this trend. The turn also to visual culture (see Evans and Hall, 1999) is a departure but also a renewal of a regard for the image which had enabled cultural studies to significantly redefine media studies.

Medium theory also has a number of recent theorists who take an extreme view of medium-as-environment. Arthur Kroker (2001) exemplifies an excessive position in which 'new media … are seen to be "new nature"', but is confused as to whether it is a metaphor or an environment. The take on McLuhan is that 'technology is an "extension" of biology: the expansion of the electronic media as the "metaphor" or "environment" of twentieth-century experience implies that, for the first time, the central nervous system itself has been exteriorized. It is our plight to be processed through the technological simulacrum; to participate intensively and integrally in a "technostructure" which is nothing but a vast simulation and "amplification" of the bodily senses' (57).

Similarly, Paul Virilio is well known for his exaggeration of a version of medium theory in recent writings on communication. In his case 'new nature' does not compete with the old, but substitutes it.

I think the infosphere – the sphere of information – is going to impose itself on the geosphere. We are going to be living in a reduced world. The capacity of interactivity is going to reduce the world, real space to nearly nothing. Therefore, in the near future, people will have a feeling of being enclosed in a small, confined, environment. In fact, there is already a speed pollution which reduces the world to nothing. (Virilio, 1998: 21)

Recasting broadcast in terms of medium theory

By far the strongest case for medium theory is the way in which we can reappraise broadcast media in a consolidated fashion. Just as, McLuhan argues, old media become the content of a new medium, so too the mediums that are made possible by old media can be viewed in a new way. Human-made environments remain unperceived by individuals 'during the period of their innovation. When they have been superseded by other environments, they tend to become visible' (McLuhan and Fiore, 2001: 17).

From the vantage point of second-generation medium theory, we can also look at some of the older media theorists and recast their work as 'medium theories' of broadcast media. This is what I attempted in Chapter 3 in explaining Althusser's theory of ideology, Debord's account of the media spectacle, Baudrillard's account of the simulacrum and the later accounts of audience studies each as versions of medium theory.

However, as we shall see in the next chapter, second-generation media theory needs to be distinguished from the fact that 'transmission' and content views of New Media have also been renewed in contemporary analysis. The study of language on the Internet, of cues-filtered-out approaches and of interaction views of CMC are each examples of such perspectives (see Table 4.2). Similarly, audience studies continues to grow as a field of media studies, but now with a renewed emphasis on the fragmented audience as well as the populist but flawed notion that the Internet commands an 'audience'.

Both broadcast and network forms of communicative action can be studied from the point of view of their content or their medium quality. In the next chapter we will see how what is common to all content (transmission) views is that they take as their building block *face-to-face interaction* and assess all communication, no matter how abstract, according to how successfully it reproduces the features of such interaction.[27]

Transmission theories are, as suggested in the earlier discussion of *extension*, only interested in how a medium may enable the continuation of face-to-face kinds of cognitive communication across time and space. They are not interested in how such extension *introduces entirely new qualities* which are not possible within face-to-face communication, a property that is common to both broadcast and interactive technology. The latter is the specialization of medium theory.

This basic distinction between the two methodologies is also related to divergent perceptions of the function of communication in social life. For the content theorists, it is cognitive interaction, but for the medium theorists, it is increasingly a matter of social integration by way of media *rituals*. The turn to ritual communication is central to second-generation medium theory. As we shall see, in media societies, attachment to mediums can be much more powerful that attachment to other people. Indeed, as a basis for community, these mediums will always be there, from cradle to grave, whilst other relationships may appear and disappear many times over.

Table 4.2 Medium theory as applied to network and
(retrospectively) to broadcast communication

	Broadcast (media studies)	Network (cyberstudies)
Content theory Transmission views (Interaction)	Information theory 1 (user, content, control, effects tradition) Shannon, Gerbner, Lasswell, Katz Audience studies 1 Mass–elite frameworks Frankfurt School – culture industry 1 Orthodox Marxist theories of ideology 1 Semiotic accounts of communication	CMC perspectives (communicative efficiency, CMC and 'perfect knowledge') Cues-filtered-out approaches Information theory
Medium theory Ritual views (Integration)	Frankfurt School – culture industry 2 Neo-Marxist theories of ideology 2 Society of the spectacle The theory of simulacra 'Medium' theory 1 Post-Saussurian perspectives The mediumization of audience studies – 'soap communities'	Virtuals community perspectives Virtual space perspectives CMC as cyberspace Medium theory 2
	Broadcast also facilitates a 'virtual community'	Sociality with objects

Communication theory cannot confine itself to the study of the inter-action between media producers and media audiences, between message producers and message receivers. Understanding the nature of communication mediums requires an understanding of communicative *integration* – the phenomenon explored in the following chapter. Even when we are not interacting with others 'through' these mediums, the mediums themselves still frame our lives.

Notes

1 Similarly, the significance of vaudeville as an entertainment form which developed a very large pre-electronic mass should be noted here, particularly for the formation of highly visible national stars (see Snyder, 1994).

2 Which typically point to a marked decline in face-to-face networks (see Guest and Wierzbiki, 1999).

3   Another example is the large number of people who participate in reader-to-reader newspaper forums (see Schultz, 2000: 214–17).

4   On this count in particular, the claims of the cyber-utopians that cyberspace can restore community and the public sphere are exaggerated. Graham and Aurigi (1998) have argued that the claims that public space has disappeared in cities are exaggerated for the purposes of this narrative of redemption. 'Not all urban trends everywhere can be generalized from Los Angeles, or other supposedly "paradigmatic" examples' (59).

5   As Geoff Sharp (1993) argues, 'technologically extended forms also stand on their own feet ... they have positive characteristics of their own' (233).

6   The scale of advertising, its budgets, the enlargement of the places in which it can occur. The contrast I am drawing here is between the dominance of the commercial dynamics of broadcasting in the first media age and the post-advertising world of user-pays communication. The contrast does not cohere in relation to public service broadcasting institutions, which, as Williams (1974) argues, rest on a paternalistic basis 'an authoritarian system with a conscience ... with values and purposes beyond the maintenance of its own power' (131).

7   In Culture Jam Lasn (2000) provides a summary rate card for 30-second advertisements on US TV circa 2000. On a national scale the Superbowl is priced at US $1,500,000 for 30 seconds; CBS news, $55,000; MTV, $4,100, $3,000. On local networks, late evening news attracts $750; Saturday morning cartoons, $450; and late night movies, $100.

8   For the greater difficulty of the Internet than radio or television as an advertising medium, see Black (2001: 402).

9   As Ien Ang (1996) documents, in early 1990 Walt Disney Studies began prohibiting cinema theatres in the USA from showing advertisements before Disney-produced movies were screened. 'The decision was made because the company had received a great number of complaints from spectators who did not want to be bothered by advertising after having paid $7.50 for seeing a film, leading the company to conclude that commercials "are an unwelcome intrusion" into the filmgoing experience' (53).

10   Studies of television have steadily emerged into a sub-discipline since their inception. Newcomb's critical readers have carried five editions since the 1970s. John Fiske and John Hartley made numerous attempts at developing a distinct television 'theory' in the late 1970s and 1980s (see Fiske, 1987; Fiske and Hartley, 1978), but by the 1990s (e.g. Hartley, 1992a) it had attained a high formalism.

11   In fact Marc argues that broadcasting as an industry, rather than a mode of integration, is returning to whence it came: '... it is easy to forget that radio emerged from the laboratory as a wholesaler's market-specific product. First known as the wireless telegraph or radiotelegraph, it was primarily sold as a wholesale military-industrial tool that extended the capabilities of telegraphy to ocean-going vessels' (632).

12   Television forms of datacasting may include: live data mining – the consumer interacts with the data synchronously with their transmission; off-line data mining – the consumer interacts with the set-top box or receiver that stores and/or updates data which have been previously transmitted; return path interactivity – the provision of a return path (e.g. a modem) which allows the consumer to interact beyond the data provided, which may include email services or on-line shopping.

13   For example, the followers of JennyCam underwent its largest growth when television programmes reporting it began to proliferate. The same is true of all privately generated sites which have become well known.

14   However, see Caldwell (1995) on the 'myth of liveness'.

15   Baudrillard's appreciation of this quality of broadcast is invaluable in pointing out the limitations of cyberactivism and the pirates of 'counter-information'. Anti-media that do not have access to the dominant forms of broadcasting are generally subject to them, and are consigned to limited expressions of 'culture-jamming'.

16   In the case of a spectacle or a catastrophe, this awareness intensifies (see Couldry, 2003: 7).

17   The modern instantaneous quality of 'news' itself is, it should be pointed out, already based on electrical technology (see Marc, 2000: 630).

18 But of course, this performativity was also only made possible because of the historical accumulation of the image of Princess Diana, the most photographed woman in world history, which had transformed her into an icon.

19 According to Robert Nisbet (1970):

> The mass is a large aggregate of people, which may or may not be in physical union, in which the unifying force is some single, usually simple, interest or idea. The television public is a mass, the crowd at a football game is a mass; the people brought together in part of a city by an incident, or report of an incident, are a mass. … It is the essence of the mass, as Simmel observed, that it is built around a single interest or aim, one involving but a part of the individual's whole nature, and that it is animated or guided by only *simple* ideas. (94)

20 On the continued popularity of the regular networks in the USA, see Buzzard (2003: 207) and Dizard (2000: 82). According to Meyrowitz, in 1990, when cable TV was experiencing a crest in the USA, 'cable households spent more than half their viewing time watching "regular" network programming and … the most frequent use of VCR's is for time shifting of programs broadcast for network-affiliated stations' (Meyrowitz, 1990: 467).

21 However, in their discussion of this function, they fail to point out how this characteristic is also endemic to radio and newsprint.

22 See the collection by Newcomb (2000) for a sample of the everyday use of TV as medium.

23 Marc claims that 'broadcasting' as an industry in the USA can be divided into radio (c. 1925–55) and television (c. 1955–85) in which '[e]ach survived the brief golden age by reconditioning its programming to supplement, complement, and otherwise accommodate the new medium that was eclipsing it' (637).

24 The closest Horrocks gets to this is in suggesting that the Internet meets all of the criteria for McLuhan's posthumously published account of the tetrad, whereby the transition from one medium to another can be measured according to a limited number of questions (see Horrocks, 2001: 60, 61, 79).

25 McLuhan, as Meyrowitz acknowledges, is the principal exponent of media as environment (see McLuhan and Fiore, 1967: 26).

26 These markers for defining a medium-as-environment discussed by Meyrowitz bear a remarkable likeness to some of the indicators of social solidarity discussed by Durkheim in *The Division of Labour in Society* (1984). Intensity, rigidity, Elements 1 and 3, underlie two dominant bases for how community tends to be discussed, by either recognition or reciprocity.

27 Even medium views are sometimes accused of this. For example, Horrocks (2001) claims that:

> Arguably, in McLuhan's paradigm, even the medium of television relies on this same metaphysical ghost of presence. Indeed, to go further, we could say with Derrida that the virtual reality environment is itself crucially dependent on maintaining presence and the immediacy of speech (the virtual discourse insists on the claim "you are *really* there"). (30)

FIVE

INTERACTION VERSUS INTEGRATION

In this chapter I will explore the difference between modern, extended forms of communication as forms of 'social *interaction*' and as forms of 'social *integration*'. To do this it is necessary to also distinguish between 'transport' views of communication, whose interest is in *interaction,* and 'ritual' views of communication, which are interested in communication as the basis of a form of community or *social integration* such as a virtual (Internet) or audience (broadcast) community.

The philosophy of communication which underwrites the transport view will be assessed and the range of kinds of interaction which this view conjectures will be surveyed. Thereafter, the idea of 'mediation' – that media are distinguishable by the fact that they are said to *mediate interaction* – is investigated. This view, which is a sophisticated extension of the 'transport' view, is finally shown to be limited in that it does not account for the fact that, as argued, some functions of communication do not require 'interaction' in the transport sense, but have the function, which defines them, of maintaining some or other level of social integration without interaction in any form.

The important sociological difference between interaction and integration is, at this point, delineated by showing that in media/information societies there are levels of integration in which 'interaction' does not always presuppose the mutual engagement of social actors, such as 'interaction' with communication technologies themselves, or 'interaction' with mediums. The abstract forms of such levels of social integration are then related to the foundational distinction between broadcast and interactivity as forms of integration which enable two distinct kinds of telecommunity, one based on *community as practice,* and another on *community as recognition.*

The strength of the integration model of understanding communication will be tested through a critique of some recent discourses surrounding cyberspace and virtual community.

Transmission versus ritual views of communication

The renewed interest in medium theory since the inception of Internet communication has come to challenge predominant transmission or 'transport' models of communication as applied to communication relationships (one-to-one, one-to-many, many-to-many) as well as individual communication technologies, from print to radio and TV and the burgeoning proliferation of New Media.

Medium theory's recovery has also been accompanied by a growing adoption of ritual accounts of communication. Ritual views of communication contend that individuals exchange understandings not out of self-interest nor for the accumulation of information but from a need for communion, commonality and fraternity. Ritual views may be micro or macro. The micro-view looks at individual attachments to objects, and the rituals people have with them, whereas the macro-view examines attachment to the mediums which these objects provide access to. As we will see in Chapter 6, in media societies, the micro-level attachment to media technologies is of an intensity which far exceeds other kinds of attachment, including with other human beings. However, at the macro level, attachment to social life and community becomes realized through these mediums – as virtual communities or broadcast/audience communities.

What is common to all ritual views, is that they suggest that mediums are not 'used' for the purposes of social interaction, but are, instead, forms of social integration. Each medium, conceived either as a technical environment or as a form of social connection, is able to facilitate a sense of belonging, security and community, even if individuals are not actually directly interacting in them. When we leave a television on in the background even when we are not watching it, or download our electronic mail when at work before engaging in face-to-face contact, we are immersing ourselves in forms of media integration.[1]

However, as pointed out in the previous chapter, the overwhelming view of the communication medium is as a conduit or vessel for the transmission of sense, of content. This is overwhelming both in the everyday common-sense understanding of communication as either 'information' or 'entertainment', and in research models into media, both old and new (see Table 4.1).

In Chapter 3, we saw how audience effects analysis, hypodermic needle models, media impact theories and computer-mediated communication approaches are each based on a transport view of communication. But what is common to all transport views of communication is a predominant set of assumptions about the nature of 'subjectivity' or identity that is involved in a communication process, and the nature of the 'messages' and meaning that are said to circulate between them. The pervasive commitment to these assumptions, as it has evolved in Western traditions of philosophy and in communication theory, is described by the philosopher Jacques Derrida as 'logocentrism'.

Logocentric communication

'Logocentrism' is a designation that has been developed by Derrida to characterize how communication in the Western world has always been marked by an over-powering desire for self-presence, the here and the now. This usually involves a quest for lost origins – a master word or signified thought, a *logos* which would transcend the uncertainties which may unsettle ordinary language. Logocentrism involves an 'exigent, powerful, systematic and irrepressible desire for a transcendental signified' (Derrida, 1976: 49). A transcendental signified functions like a primordial self-presence, which is somehow independent of the play of signs, and which would bring an end to that infinite regress for a final meaning in the certitude of a ground. It lies at a point where representation withdraws into a transparent unity with what it represents, where subject and object are joined at last *in presence* in a privileged connection to truth and meaning.

For Derrida, logocentrism is inscribed in a range of features of Western thought and culture, including an imperative to decide quickly, to define who is foe and who is friend, anxiety over having to have certainty and truth, the desire for instantaneous and perpetual contact – a supernormative impulse for self-presence and certitude. For the purposes of this book I am primarily going to detail the implications of logocentrism for *communication*.

In an early collection of his most accessible interviews, *Positions*, Derrida (1981) outlines the main features of logocentric communication as the communication of consciousnesses. The logocentric concept of communication involves 'a *transmission charged with making pass, from one subject to another, the identity* of a *signified* object' (23).

This conception also borrows from the value of presence involved in auto-affection or 'phonocentrism', where the temporal and spatial are united in a way which provides an increased measure of guarantee that a message will reach its destination: a situation in which the speaker hears him- or herself speak at the same moment as the hearer does. In this model of communication, subjects are posited as the self-present symmetrical poles of an intersubjective process. The other value which is central to logocentrism, as we have seen, is the view that there exists an inventory of fixed signifieds which precede and are anterior to the speaking subject and which are merely drawn upon in order to communicate meaning and make present a *common* reality. Against the logocentric concept of communication, Derrida argues, as we shall see, that *language is constituted in the very difference and distance that communication in its semiolinguistic sense is supposed to overcome.*

The metaphysics of the sign

Derrida's critique of the logocentric concept of communication is at its clearest in his account of Saussure.[2] Saussure was one of the first thinkers to explore the stability of the relationship between signifiers (marks on a page, sounds in the air) and signifieds (thought concepts). When these two elements are associated, they become a sign. Saussure showed that it is the relationship not between signifier and signified but between systems of signifiers which determines the kinds of thought concepts that are associated with any one signifier.

Because of this, the relationship between signifier and signified is not 'fixed for all time' but is in fact arbitrary, even if it may be fixed by social convention. The consequences of this claim are far-reaching.

To begin with, it means that language is a system without positive terms. There are no self-present terms with which one can anchor other terms.[3] Saussure (1922) proposed that 'whatever distinguishes one element from another constitutes it' (118–21). To repeat a signifier it is not necessarily true that a signified associated with it in a former signification will also be repeated. In this, Saussure, as Derrida would be the first to tell us, contributed powerfully in overturning the metaphysical notions of language as a nomenclature or a naming system.

However, Derrida's criticism of Saussure is directed first at the notion that there is such a thing as a *stable* sign 'system' as a kind of general context and, secondly, at the notion that the signs which constitute the system are full enough in their identity in order that we may discern differences between them. Thirdly, there is a problem of agency insofar as Saussure seems to be assuming that subjects do actually discern differences between signs in a more or less uncomplicated way. What is problematic in Saussure is that he regarded the relation between a signifier and a signified to be in a cosy one-to-one correspondence, as in a parallelism, as if all signs were constituted by symmetrical values (that could be measurable) of the signifier and signified.

Derrida's critique is that there is no such thing as a 'closed', or what he calls 'saturable', context of meaning, and that signs somehow 'possess' a fullness of meaning (a plenitude) by which they are differentiated from other 'full' signs.

Communication and dissemination

This critique is well set out in an important article which formalizes Derrida's thoughts on the topic of communication: 'Signature, Event, Context' (hereafter SEC, Derrida, 1986). In this article, a sustained analysis is developed.[4] In particular, Derrida addresses the question of 'contexts' of communication. He proposes to demonstrate that there is no such thing as completely saturated or homogeneous contexts, which would have two consequences:

- 'a marking of the theoretical insufficiency of the *usual concept of context* [the linguistic or non-linguistic]' (310);
- 'a rendering necessary of a certain generalization and a certain displacement of the concept of writing ... which could no longer ... be included in the category of communication ... understood in the

restricted sense of the transmission of meaning. Conversely, it is within the general field of writing thus defined that the effects of semantic communication will be able to be determined as particular, secondary, inscribed, supplementary effects' (SEC: 310–11). Derrida says that a new concept of writing is bound to intervene which will transform itself and the problematic, i.e. transform the problematic of thinking communication on the basis of the semantic and the non-semantic and the system of interpretation which is hermeneutics (see SEC: 309–10).

In SEC and in other texts, Derrida advances a new concept of *writing* whose purpose is to overturn 'a definition of language as *communication*, in the sense of the communication of a *content*' (1988: 79). Derrida attempts to demonstrate that the effects of semantic communication are subordinate (both *de facto* and *de jure*) to the effects of *writing – defined as the impossibility of an homogeneous context* – which can simply be enhanced to varying degrees by technically more *powerful* degrees of mediation (such as the vulgar 'classical' concept of writing). This force of writing throws into confusion the non-semiolinguistic concept of communication, which carries with it the sense of bridging a gap or opening an aperture.[5]

Derrida redeploys 'writing' in a special way not simply as a label for words on a page (i.e. a technical medium) but as a term which he opposes to the way in which speech is thought of as carrying the *logos*. Rather than suggest that writing is merely an extension of speech, he reverses this claim by showing that speech, the ability to produce meaning, is governed by certain properties that have been historically recognized in the conventional notion of writing. In particular, there is a force in writing that he calls *dissemination*, or the way in which meaning is never self-present or 'full' but is always escaping to numerous other contexts, as well as being borrowed from other contexts over time and in space.

Derrida reduces the effects of language to two – polysemia and dissemination. In SEC Derrida speaks of 'the necessity of, in a way, *separating* the concept of polysemia from the concept I have elsewhere named *dissemination*, which is also the concept of writing' (SEC: 316).

> The semantic horizon which habitually governs the notion of communication is exceeded or punctured by the intervention of writing, that is, of a *dissemination* which cannot be reduced to a *polysemia*. Writing is read, and 'in the last analysis' does not give rise to a hermeneutic deciphering, to the decoding of a meaning or truth. (SEC: 329)

Dissemination is a force of rupture which is not reducible to 'the horizon of a dialectics', to 'the work of the negative in the service of meaning' (SEC: 317).

Dissemination is defined as one of the sides (or effects) of the iterability (repeatability) of the signifier (a word on a page, a sound in the air, which Derrida also calls a 'mark'), the side in which the mark fails to reproduce a signified content. The ability of a mark to reproduce a signified content can be reduced to polysemia, the production of a range of possible meaning-effects which are the result of the contextual chains in which the mark can be associated in a horizontal fashion.

For Derrida, polysemia is a rather modest aspect of signification, a 'narrowly semiotic aspect' of his concept of writing (1981: 29). Signs are able to signify things, but what is neglected is that they do so in the face of the alterity of the sign, the fact that it plays a role elsewhere, in other contexts at other times.

The difference between polysemia and dissemination has nothing to do with the *size* of contexts; rather, they are two opposing forces implicit to all contexts, and scale is irrelevant. Nor does dissemination have anything to do with the relation between unmediated and mediated communication. The nature of mediation is that it implicitly presupposes that there is some substance that is being mediated. It cuts in always *before* (always-already before) the effects of representation and a semantic field. Its effects are precisely those of a structural indifference towards performative communication and the impossibility of saturation. The more information that one has about a context, the less successful will be the possibility of totalizing the interpretation of utterances.

For this reason, there can be no neat narrative about what individuals think with regard to the failure or otherwise of the sign to signify. Dissemination does not make its effects felt in the service of meaning; rather, the phenomenal level of meaning is materially foreclosed by the way that an audience, a viewer, a listener, a reader, is caught up in dissemination.

The effects of dissemination ensure that 'there can never be an experience of *pure* presence, but only chains of differential marks ... because the iterability which constitutes their identity never permits them to be a unity of self-identity' (SEC: 318). The consequence of this is not to say that 'the mark is valid outside its context, but on the contrary that there are only contexts without any centre of absolute anchoring' (SEC: 320). When the statement is made, 'we need to put a matter "in context" or "in perspective"', a logocentric promotion to the idea of a stable context is being reinstated.

The difference between dissemination and polysemia, then, inheres in the latter's being determined by, and functioning within, the 'imaginary' experience of an homogeneous context, even if this context is as abstract as that which receives its definition from an *arche* (a starting point) or *eschaton* (an end point) – as in the philosophy of Hegel. Within this context the task of interpretation is still to recapture the possible meanings within the text – by means of a totalizing dialectics, for example. Derrida's critique of polysemic interpretation is precisely to deal with the desire for totalization (the fallacy that contexts are finite and totally knowable) and self-presence (the dream of a sign which would refer to nothing but itself).

Transcendentally, totalization and self-presence are both impossible, then, because all contexts are inhabited by radical alterity or *dissemination*. Moreover, Derrida argues that today, in the era of telecommunication, we have become more attuned to these realities of language-as-writing. The ideology of telecommunication is that, in a non-semiolinguistic sense, it aims to 'bring subjects together'. But, argues Derrida, the purposive rationale of telecommunication conceived as the transport of semantic communication is *de jure* undermined at its very origin. Yet, on the other hand, the fact of telecommunication as a non-semiolinguistic phenomenon, as that which enables the citational grafting and lifting of marks out of local contexts, that is, as the material basis of that always open possibility, establishes *de facto* a form of subjectivity which is constitutively more abstract than social forms which relied on the illusion of closed contexts. That is, the techno-utopia of global-village discourses proposes a return to the immediacy of social relations whilst in fact it dissolves those relations in its very production.

Writing and telecommunication

What writing and telecommunication do in relation to 'the transparency and immediacy of social relations' is to introduce the always open potential for a meaning to be abstracted from its 'original' context, in the way that it is experienced as original. In other words, this is to say that the distinction between the original and the repeated becomes entirely open and indeterminable; this is how writing extends and abstracts individuals from closed contexts (see SEC: 320). What Derrida celebrates in the modern period is the more transparent move towards social relations assuming a form in which discourses refer to nothing except themselves (as in a simulacrum); to no grand narrative or originating discourse, even though there may be a nostalgia for such things. As such a condition, which reveals the *disseminative* side of writing, begins to prevail, the more 'local contexts come into contact owing to more powerful and universal means of communication that can be found in the era of telecommunication and the Internet. These developments in the means of communication de-parochialize contexts by bringing them into contact with others. That is, the possibility of reproducing a meaning (what Derrida calls the 'internal context' in the semiolinguistic sense: SEC: 317) is undermined by the very practice of its reproduction the more a mark is mass-reproduced (but *not* in the sense of permanence). The efforts to intensify mass communications can only create greater abstraction. So writing, which, *de jure* is normally thought of as a technology for the reproduction of meaning, when considered logocentrically (the semiolinguistic sense) is also the medium which *de facto* destroys this very ideal as lived wherever it operates.

In theorizing the rise of telecommunication, Derrida argues against the idea of a 'global village', as popularly conceived as a universal community. The 'network society' is marked not by a change in semiolinguistic extension (an ability to extend meaning) which works the same for both speech and writing, but by the phenomenon of *language-as-writing* as a technological power in itself changing and producing more and more non-saturable and open contexts.

The notion that 'writing'overcomes the problem of immediacy is as ideological as the notion that this already occurs in relations of mutual presence or of performative, event-like statements. And yet Derrida is enthusiastic about telecommunication not because it creates a global village but because it undermines the ideological notion of an homogeneous context by installing the material power of writing as a non-saturable context. The kind of threat which only writing makes possible to logocentrism is now made more universal than ever following the rise of telecommunication.

> As writing, communication, if one insists upon maintaining the word, is not the means of transport of sense, the exchange of intentions and meanings, the discourse and 'communication of consciousnesses'. We are not witnessing an end of writing which, to follow McLuhan's ideological representation, would restore a transparency or immediacy of social relations; but indeed a more and more powerful historical unfolding of a general writing of which the system of speech, consciousness, meaning, presence, truth, etc., would only be an effect. (SEC: 329)

Derrida is here critical of how McLuhan singles out writing (in the narrow sense as defined by Derrida) as the one technology of communication which threatens the, by his account, sensory richness of an oral culture by means of an over-dependence on vision. The emergence and predominance of exclusively electric technologies are posited to facilitate the restoration of sensory balance (especially the liberation of the said improvisory, gestural and synaesthetic qualities of speech), counterbalancing the linear and mechanical culture of what McLuhan (1967) calls Gutenberg or typographic man.

For Derrida, the effect of privileging the narrow definition of writing is that it reinstates a dichotomy between speech and writing which in turn reinstates the phonocentric metaphysics of presence: 'It is this questioned effect that I have elsewhere called *logocentrism*' (SEC: 329). The 'powerful historical unfolding of a general writing' arrives to release the radical potential of the 'essential predicates' of the general, disseminative status of writing which have always been repressed and denied by logocentrism:

> ... writing, as a classical concept, carries with it predicates ... whose force of generality, generalization, and generativity find themselves liberated, grafted onto a 'new' concept of writing which also corresponds to whatever always has resisted the former organization of forces, which has always constituted the remainder irreducible to the dominant force which organized the – to say it quickly – logocentric hierarchy. (SEC 329–30)

Experiencing mediums the logocentric way

The logocentric metaphysic of communication is one that permeates most of the Western experience of all of the mediums that are discussed in this book – whether in terms of individual communicative technologies or of communicative architectures.

In much communication theory too, many of the various assumptions which underpin logocentrism can be seen to operate. These are: a philosophy of the medium as vessel, of individuals as 'users' of this kind of medium, and of meaning as the product of intentionality. Two mediums are postulated within logocentrism: firstly, *natural language* as a conduit; and, secondly, a *technical means* of conveying language, such as print, television or on-line communication networks.

It follows that, within the logocentric tradition, some technical mediums are viewed as 'conveying' messages more powerfully than others, and are able to provide an immediacy that others are not. The ideology of such a state of communication is virtual reality itself, in which the process of representation withdraws to the point where only the represented remains. The yearning for such a condition of unmediated transparency is, as Bolter and Grusin (1999) argue, especially pronounced in the context of digital technology. They claim that, from the early 1990s, new digital media, together with the way older media have remediated to take on the form of the new, 'fulfill our apparently insatiable desire for immediacy' (5).

> Live 'point-of-view' television programs show viewers what it is like to accompany a police officer on a dangerous raid or to be a skydiver or a race car driver hurtling through space. Filmmakers routinely spend tens of millions of dollars to film on location or to recreate period costumes and place in order to make their viewers feel as if they were 'really' there. … In all these cases, the logic of immediacy dictates that the medium itself should disappear and leave us in the presence of the thing represented: sitting in the race car or standing on a mountain top. (5–6)

Bolter and Grusin argue that immediacy depends on what they call 'hypermediacy' – a fixation with the medium itself (6). Where 'one medium seems to have convinced viewers of its immediacy, other media try to appropriate that conviction' (9). They point out that increasingly during the 1990s televised newscasts, with their multiple panels of text, image and logos, 'came to resemble web pages in their hypermediacy' (9). These examples of multi-media enhancement of the television medium are driven not simply by demands for rich formats of information, but by a metaphysical commitment to expressivism, and the transmission model of communication.

As suggested by Derrida, the experience of medium in logocentric culture is predominantly an instrumental affair. A medium is lived as just a tool for expressing meaning, just as language is viewed as a transparent medium for transporting sense. Such a medium is conceived as having authors who are autonomous, unified meaning-creators, who must draw upon the stock of meanings available to them to produce symbolic forms. The meanings which can be conveyed in a given medium are regarded as stable, and antedate the actual communication event. Communication is the transport of these stable intended meanings: the communication of consciousnesses. The task of readers, listeners or viewers of texts is to understand what the author intended to say – what Derrida calls 'hermeneutic deciphering'.

In pre-Saussurian accounts of communication this meant that to repeat a signifier is to repeat the meaning or the concept that supposedly accompanies it. The task of the author is to select the appropriate signifiers: 'I am searching for the word to convey what I mean.' For the reader or audience, meanwhile, communication is completed by identifying the signifieds which the author is seen to have attached to the chosen signifiers. This notion of correspondence also implies that it might be possible for a signifier to fail to carry the signified that should accompany it. This logocentric system implies that it is somehow possible to re-create the 'original context' in which a signifier had meaning for an author, or producer of symbolic forms.

However, for Derrida, as we have seen, the meaning of a signifier at the point in which it is consumed will always be different from its meaning at the point of production. The so-called 'original context' can never be reproduced.[6] Moreover, signifiers do not even exhibit stability within the *same* context, and may be read by different audiences in a diverse number of ways.

The open-endedness of *writing-as-language* ensures that there is no such thing as understanding an author 'out of context', because there is never an 'original' context which can be captured by a logocentric reading; rather, the context is the reading, the translation, etc.

For Derrida, we would be searching in vain to find an original context or an original author, just as we would be to yearn for an original 'meaning' or, perhaps, as positivist philosophers do, search for the 'meaning of meaning'. It is radical alterity which ensures that the one-to-one correspondence between a signifier and a signified, intentionality and the 'original context', is a myth, a theological idea.

Nevertheless, Derrida would be one of the first to acknowledge that the logocentric mode of experiencing communication is a pervasive one in Western societies, and one that looks set to survive the de-parochialization that accompanies new means of global communication.

Experiencing mediums as sites of ritual

In keeping with the logocentric metaphysics of communication, transmission accounts take dyadic interaction as their building block of analysis, and then examine more and more complex forms of mediation that alter this building block. Transmission accounts do not ignore the kinds of mediation that are in play in a communication process; indeed, they are interested in how these mediations distort or inhibit the continuation of dyadic interaction as a way of appreciating what they are.

However, as Meyrowitz points out, the idea of *intentionality*, which is the most dominant metaphoric framework by which mediums are 'lived', cuts across all kinds of communication. We saw in the previous chapter that the metaphysics of *intentionality* suggests that the *message is the medium*. And, within this metaphysics, the more extended across time and space are the mediums by which we can express intentionality, the more convincing it is to live by the instrumental metaphor of communication. The greater the distance that is covered in a communicative event, the more assured we can be of the validity of logocentric communication.

We have already seen how the 'metaphysics of presence' which characterizes the Western experience of communication is premised on the phonocentric analogue of face-to-face exchange. This does not mean that the *act* of face-to-face communication is preferred over other kinds of communicative acts; rather, all forms of communication tend to be metaphorically lived and evaluated in terms of face-to-face communication.

For example, the vast critical literature bent on discrediting the social utility of television, which had its origins well before the Internet, is largely aimed at its inadequacy for face-to-face-like interaction.[7] Likewise, the Internet itself is also assessed for what qualities of face-to-face communication it can redeem by simulation. Where a medium is deemed to offer more opportunities for such simulation, those interacting with mediums are typically described as 'users'. Thus, for example, consumers of broadcast technologies are seldom called 'users', whereas those who interact with email, the World Wide Web or interactive TV frequently receive such a designation.

To think of technological mediums in this way is to follow a familiar fallacy that equates communication, transportation and exchange systems as 'service functions' of culture, as supplements to social reproduction. By this view, individuals are regarded as tool-using actors who are placed at either end of intersubjective processes. In such a view, the message is the medium and the medium is 'used' as a tool.

In contrast to this model, the advent of Internet communication has coincided with a renewed interest in theories of ritual communication. What theorists of the Internet have taught us is the importance of looking at the ritual aspects of technologically extended communication, which also means looking at the distinction between interaction and the way in which communicative forms carry social integration. Sometimes, this connection is only intuitively observed, as when, for example, Howard Rheingold (1994) describes communication on his own Internet network, the WELL, as a 'bloodless technological ritual'.

Part of the reason why ritual views have become popular is that cyberspace is so much more visibly a medium than is broadcast. Both the medium and the content of cyberspace can be analysed, but the content is not as socially visible as is the content of broadcast. I will return to this contrast in the sections below on the nature of 'interaction' in broadcast and network architectures.

As the medium of broadcast is less visible, media studies analysis has always tended to gravitate to messages and genres, ownership and control. But an important counter-tradition to this can be found in the work of James Carey and Elihu Katz.

Carey's work has attained considerable popularity as a kind of modern-day foundational expositor of the ritual approach, his key text being *Communication as Culture* (1989). For Carey, social reality is partly *produced* in communication, not simply reflected by it. Quoting from John Dewey, Carey argues, 'Society exists not only by transmission, by communication, but it may fairly be said to exist in transmission, in communication' (14).

Media might reflect the world, but they also produce the world. And when individuals interact with media, they are, as Durkheim would suggest, worshipping the way 'society substitutes for the world revealed to our senses a different world that is a projection of the ideals created by the community' (Carey, 1989: 19). However, such a ceremony need not be driven by the desire for a 'metaphysic of presence', since typically the medium that is interacted with is invisible.

Carey's view of media is, in a way, remarkably similar to that of Derrida and Baudrillard, in elevating the importance of the performative role of texts, whether these be media texts or written texts. As Carey conjectures, 'The particular miracle we perform daily and hourly – the miracle of producing reality and then living within and under the fact of our own production – rests upon a particular quality of symbols: their ability to be both representations "of" and "for" reality' (29).

Carey's formula can be seen to parallel Derrida's claim about the double dimension of the sign: one aspect of it concerns the reproducibility of meaning, whilst the other is about the fact that texts themselves constitute their own reality, which refer not to a beyond, but to each other, to intertextuality, which we can only gain access to via specific mediums.

In mounting the argument for communication as ritual, news is a central focus for Carey because it is so typically thought of in terms of information alone:

> If one examines a newspaper under a transmission view of communication, one sees the medium as an instrument for disseminating news and knowledge, sometimes *divertissement*, in larger and larger packages over greater distances. Questions arise as to the effects of this on audiences, news as enlightening or obscuring reality, as changing or hardening attitudes, as breeding credibility or doubt. ...

> A ritual view of communication will focus on a different range of problems in examining a newspaper less as sending or gaining information and more as attending a mass, a situation in which nothing new is learned but in which a particular view of the world is portrayed and confirmed. (20)

Under a ritual view, 'news is not information but drama' (see Carey, 1989: 21; Morse, 1998: 36–67). But this drama need not be simply about communion but can be based on anxiety. News is a premier genre for invoking such anxieties, and the addiction of audiences to tune in to a daily update can be to ritually satisfy Dionysian needs (See Alexander, 1986)[8]. However, from a transport point of view, individuals are viewed as simply 'using' the media to overcome their anxieties, rather than the media having produced the behaviour which is supposed to be overcome.

The transmission model of communication, on the other hand, promotes the illusion that messages comprise 'information' that is simply *learned* and that they externally influence otherwise unmediated behaviour. Such a view takes no account of our attraction to such media in the first place, and the way in which a 'media event' can come to mean as much, in terms of attachment, or more, to individuals as non-media events.

Carey is also critical of prevailing views of the power of media: 'We are (mistakenly) coerced into thinking of communication only as a "network of power" which needs to be "balanced" at the level of content in order to legitimately represent the pluralist interests of liberal democratic societies' (34). Such power derives not from media influence over consciousness as hegemonic networks, but from a form of remapping and displacement of primary environments of recognition and identification.

Both the ritual and the power dimensions of media have been in Carcy's view obscured by 'uses and gratifications analysis' (32). The 'uses' model reaffirms the notion that media are an instrumental extension of processes of moral development (18). From telegraph to computer, communication is seen to be about moral improvement. This is usually expressed as new technologies being used to carry older forms of social relationships in a new, more 'helpful' medium. This latter view is one which can be characterized as a purely informational view of communication, in which communication becomes either a means of control or a means of expressing individuality.

Table 5.1 summarizes the major differences between transmission and ritual approaches.

Now that we have examined the difference between transmission and ritual accounts of communication, I want to investigate the difference between interaction and integration in terms of various typologies which have been put forward in communication theory. We shall begin with typologies of *interaction*.

Table 5.1   Transmission and ritual perspectives compared

Transmission view	Ritual view
Concerned with content	Concerned with medium
News is information (Carey, 1989: 21)	News is drama and performance
Individuals interact with each other	Individuals interact with a medium
Logocentric – individuals restore presence	Simulacra – the act of communication does not refer beyond itself
Symbols are representations 'of' (Carey, 1989: 29)	Symbols are representations 'for'
The media 'mediate' reality	The media produce reality
Interaction	Integration
Face-to-face is privileged	Face-to-face is marginalized
Fleeting	Constant

Types of interaction

As has been suggested, the metaphoric framework of logocentrism is one which privileges interaction-as-event. Because of this, 'phonocentric' and auto-effective forms of communication are privileged over others. Phonocentric interaction exhibits a number of features which are differentially given continuation by different mediums.

These are the qualities of being 'live', and of mutual presence between two or more interactants, who each have an opportunity for speech. What these qualities provide, which can be 'lost' in extended communication, is a rich range of contextual information. Regardless of what is known of each interactant in such a setting, the fact of their presence; their body language, gestures and symbolic expressions; the way that they acknowledge other interactants with glances and expressions; brings with it a ready-made environment of meaning.

When we look at extended media, however, these two qualities, of synchronicity and mutual presence, are, by definition, no longer co-present. No extended media can reproduce all of these qualities of phonocentric communication at once. The project of virtual reality is one which dreams of such an achievement, but, paradoxically this requires making the body redundant.[9]

However, extended media are certainly capable of singling out one or two of these qualities in a typically enhanced fashion. Thus broadcast can be viewed as an exaggeration of the auto-affective features of phonocentric communication events – it isn't live merely for two people, but potentially for billions of people. However, from the standpoint of the same model, it does not measure up in allowing all participants an equal opportunity for speech. Conversely, interactive telecommunication is a very powerful means of two-way or multiple-way communication, but is seldom live, or in real time. And even when it is real time, the absence of mutual presence attenuates the deictic gestures possible in phonocentric communication events.

These basic differences between kinds of extended media, and the fact that they typically extend some aspects of phonocentric communication whilst they annul others, is well outlined by John B. Thompson in his book *The Media and Modernity* (1995).

In a sophisticated typology of interaction, Thompson distinguishes between three types of interaction: face-to-face interaction, mediated interaction and mediated quasi-interaction. 'Face-to-face interaction occurs in a *context of co-presence*; the participants in the interaction are immediately present to one another and share a common spatial-temporal reference system' (82). In the face-to-face, participants can 'use deictic expressions ("here", "now", "this", "that", etc.) and assume that they will be understood' (82). Face-to-face communication is, like some extended modes, dialogic, or two-way. However, Thompson claims it is a special dialogic, because it is the only communicative event in which participants 'are constantly and routinely engaged in *comparing* the various symbolic cues employed by speakers, using them to reduce ambiguity and to refine their understanding of the message' (83, italics mine). It is the multiplicity of symbolic queues that are seen to be available to face-to-face situations that guarantees a form of presencing which Derrida names 'phonocentrism'.

Thompson's next form of interaction is 'mediated interaction', which includes letter-writing and telephone conversations. It presupposes a technical medium (paper, electromagnetic waves, etc.) which enables messages to be transmitted to persons remote 'in space, in time, or both'. The most important feature, however, is that '[w]*hereas face-to-face interaction takes place in a context of co-presence, the participants in mediated interaction are located in contexts which are spatially and/or temporally distinct. ...* The participants do not share the same spatial-temporal reference system and cannot assume that others will understand the deictic expressions they use' (83). Because of this, communicants must decide how much contextual information to add, such as signatures, letterhead information, or identification at the start of a phone conversation.

The third form of interaction is 'mediated quasi-interaction'. This form is peculiar to the media of mass communication – books, newspapers, radio and television – and its defining feature is that 'symbolic forms are produced for an indefinite range of potential recipients' (84). This level of interaction is one which engages individuals 'impersonally', but does not exclude them from more horizontal forms of personal association. Of course, the early days of 'mass communication' and effects analysis assumed that something equivalent to this level of interaction was the dominant form of interaction of mass society – so much so that the primary group of 'face-to-face' relations had to be 'rediscovered' in this tradition (see Lowery and De Fleur, 1983: 180).

For Thompson however, it is because the addressees of this form of communication do not have specificity, *to the extent that this form engages them*, that individuals look also to primary forms of association. Nevertheless, *quasi*-interaction, even if it is largely 'one-way', is still a

Table 5.2 John B. Thompson's instrumental/mediation paradigm

Types of interaction	Qualities
Face-to-face interaction (mutually embodied presence)	Dialogic Mutual presence A high degree of contextual infomation (body language, gestures, symbolic cues, deictic expressions: 'here', 'this') Reciprocal Interpersonal specificity
Mediated interaction (technical mediums like writing, telephoning)	Dialogic Extended/not mutual Restricted degree of contextual information (letterhead, signature, date placed on communication) Reciprocal Interpersonal specificity
Mediated quasi-interaction (books, newspapers, radio, TV)	Monological Extended Produced for an indefinite range of recipients by a small number of media producers Senders and receivers of messages nevertheless form bonds

form of interaction, according to Thompson, because it 'links people' together (84).

As per the summary of Thompson in Table 5.2, the two forms of extended interaction are significant in the way they correspond to interactive versus broadcast communication. The quasi-interactive quality of broadcast is precisely the feature that second media age thinkers are critical of. Second media age thinkers are critical of the fact that there are agents who stand between a sender and receiver of messages. The now fashionable concept of 'disintermediation' that has emerged in recent literature is entirely circumscribed by this rejection of mediation-by-agents (see Dominick, 2001; Flew, 2002). This concept is primarily confined to describing the economic functions of media in connecting buyers and sellers (the removal of the 'middleperson'), but has also broadened out to the cultural functions of media. Somehow, the heightened dependence on CMC which replaces such mediation-by-agents isn't also seen to be a form of mediation (as Thompson's model proffers). Disintermediation only refers to the removal of human agents in the media process.

Oddly, machine-assisted or electronic means of communication are somehow exempt from the mediation process, as if they are transparently a means of rescuing the face-to-face from the way it suffers at the hands of mass media. However, what can be noted in Thompson's typology is that whilst these interactive forms differ, they have in common the fact that they are *both* viewed as *a mediation of face-to-face interaction* and that from this standpoint the term 'disintermediation' is flawed. There is no reason why, within the logocentric framework of 'mediation' theory, using submedia of the Internet, for example, is any less 'mediated' than watching television.

However, as we shall see, whilst it is erroneous to classify either network or broadcast dynamics as either more or less 'mediated', the conceptualization of mediation itself must be emancipated from the transmission model. Thompson's model also sits firmly within the communication-as-transport paradigm of communication – a transmission-based model in which the face-to-face is the base and is mediated by an architecture that extends it in some way.

The problem with 'mediation'

Thompson's model is useful in the way it summarizes the background theoretical architecture that is drawn on by many of those attempting to theorize old and new media. For example, the 'cues-filtered-out' approach is easier to understand in this comparative typology of media forms. Nearly all of the cyberspace literature is framed by a social interaction model – i.e. that face-to-face interaction is being supplanted by extended forms of communication – and this is seen to be derived from technology somehow intervening and separating us from some 'natural state' of interaction which is the face-to-face.

More useful still in Thompson's typology is the way in which it takes the interaction model to its outer limit. And the limit of this model is precisely the question of communicative *context*. We can recall Thompson's claim that '*the participants in mediated interaction are located in contexts which are spatially and/or temporally distinct*'. A number of observations can be made here:

- What is assumed in this statement is that the communicative *context* which is to be privileged over all others is the 'local', 'embodied' context capable of mutual presence, which is separated from other such 'local' contexts.
- The task of communication is to overcome such separation by the transmission of symbolic forms.
- Interactants have some measure of control in overcoming this separation of contexts by actively adding specific cues that are otherwise structurally absent.

Mediation theory is a variant of instrumental theory, discussed above. Instead of an interactant simply using a medium to convey a message, the meaning of the message will be affected by a range of factors that come to mediate its successful transmission. However, intentionality may be preserved insofar as the interactant understands the medium and is able to make allowances for such mediation in crafting a particular message. Thus, for example, journalists learn how to 'write across' different media. In this way, interactants are seen to 'metaphorically' extend what they may be accustomed to in face-to-face interaction. This is because they inevitably project the qualities of face-to-face communication onto the extended forms.

However, this description of mediated interaction assumes either that the 'local context' of interlocutors in technologically extended communication is already known (be this mutually or otherwise) or at least that it is in part disclosed in the communication process. The interlocutors are deemed not to have an identity outside of what is known of their local context. The local context gives them their identity. They may venture out of it and communicate with others across time and space, but they do so anchored in their context of origin, and are said to be identified in this way.

To return to Table 5.2 once more, we can see how mediated and quasi-mediated forms of interaction are seen to be continuations of the face-to-face by other means. The qualities listed beside these forms are in terms of what is replicated from the face-to-face, of what is absent. Thompson does not, however, add *what is unique to* technologically extended media which are absent in the face-to-face. Thus, for example, under mediated interaction, he could add the fact that print is capable of information storage, or, in the quasi-mediated interaction list, the fact that broadcast communication is synchronous between large audiences, something unachievable in mutual presence.

However, technologically extended communication, whilst annulling some of the features of phonocentric communication, also adds qualities which are not possible within such communicative events. The technical apparatuses of communication mediums can produce sounds not generated by the human voice, text too difficult to produce in the course of ordinary dialogue, as well as images beyond merely the image of the person engaged in interaction. These qualities, which Meyrowitz identifies (see previous chapter), can be thought of as providing the resources of an entirely different register of communication 'language' – the arrangement of images, camera angles, fades, echoes, sound-mixing genres, textual conventions and styles, etc. The way that producers and consumers of such images, music and text relate to them will vary considerably. However, despite this, all interlocutors appreciate that the meaning of messages in these mediums is to some degree shaped by the medium.

In such cases, the idea that the face-to-face is simply mediated begins to collapse as it is realized that technologically constituted mediums *bring about new contexts* of a substantially different order than can contexts capable of mutual presence. To insist that there exists a separation of contexts can only be upheld in an instrumental model of communication. In ritual views, however, there is no separation; the medium *is* the context.

But the mediation view is very entrenched. It is even built in to nomenclatures of communication like 'computer-mediated-communication'. The problem with the mediation view is that it replicates, in the field of communication studies, what is endemic to common-sense views of technology-in-general – an instrumental perspective. It does not see communication technology as substantively capable of its own context and the dependence that individuals might have on this context.

The difference between instrumental and substantive views of technology is well explored by Feenberg (1991). In Feenberg's account, the instrumental perspective 'treats technology as subservient to values established in other social spheres (e.g. politics and culture)', while substantive theory 'attributes an autonomous cultural force to technology that overrides all traditional or competing values' (5). Substantive theory sees technology as embedded in circuits of interaction, and 'argues that technology constitutes a new type of cultural system that restructures the entire world as an object of control' (5). The more technology comes to mediate the kinds of engagement and interaction individuals have with the world, the more it takes on the character of an environment rather than a tool. Moreover, technology has the power to encourage individuals to view environments in particular ways that are shaped by the means they have of relating to them.

Raymond Williams' distinction between technical invention and technology-as-socially-configured is valuable here (Williams, 1974). He suggests that in fact even in the first media age 'technology' cannot be empiricized (i.e. technology-as-object) but, in a more Heideggerian way, is always-already located as a medium (e.g. broadcasting).

This blind spot of second media age thinkers for not seeing technology in this wider sense allows them, in my view, to conflate their own critique of first media age theory with their projection of first media age 'technologies' as fundamentally tool-like. For this reason, so much is invested in the term 'cyberspace', whereas, if Williams' distinction is adopted (with the addition of a list of qualifiers about the nature of interaction in broadcast technology), first media age technology might also be considered a kind of cyberspace, serving to level further the distinction between first and second media age.

Medium theory and individuality

Like transmission accounts, medium theory typically looks at how the position of the communicants and the information communicated is determined by different media. But it also suggests something quite radical and different from transmission accounts – the possibility that individuality itself is (at least partially) an effect of a medium. In this view there are no pre-given subjects with an experience of the real. There are no transcendental contexts which pre-exist other contexts and determine how they are experienced.

In Thompson's case it is embodied, mutual contexts which are seen to pre-exist the extended contexts, and these former contexts structure how we engage with the latter contexts. Moreover, the extended contexts mediate our attempts at re-creating local contexts, even when they may be separated in space and time.

The position of the individual in these settings is very much conceived logocentrically. Intentionality, the idea of being a language-user, and even the idea of being a subject at all, is tied to this essentialism.

In contrast to this, medium theory suggests that individuality is an *effect* of the medium itself. Such an idea was really first glimpsed by Adorno and Horkheimer's account of the culture industry, which they describe as 'a circle of manipulation and retroactive need' (see Chapter 2 for a discussion).

Althusser's model of ideology-in-general is also interesting in this regard. Whilst, ostensibly, he was interested in the formation of identity in ideology as an unconscious structure, his account purveys a 'constitutive' account of subject formation. There are no pre-given subjects with an experience of social reality; rather, they are constituted by an apparatus of integration. Althusser named the mechanism of this apparatus 'ideology-in-general'. Ideology-in-particular is the name given to the content of ideology, a question of consciousness, whereas ideology-in-general describes the very operation of 'calling' or hailing individuals as subjects of a system of broadcast.

For Althusser, therefore, a particular mass media kind of identity is peculiar to the structure of broadcast, but not of interactive networks of communication. Althusser does not theorize the forms of reciprocity and identity formation produced by extended telephony or letter-writing. The Internet, too, had only become fully operational in the year that Althusser died. But network integration in general was for him an inadequate terrain for examples of interpellation. Interpellation only occurs within 'one-to-many' apparatuses.

It is only recently, in the development of the theory of the avatar, that accounts of subjectivity constituted in extended interactive mediums have also come to the fore. In this view, the medium-as-environment does not simply hijack a former kind of identity, but constitutes a new kind of identity which lives a kind of virtual life. The TV channel surfer, the Internet avatar, the writer and reader immersed in the text, etc., are each an example of this relationship. Implicitly, therefore, medium theory espouses an account of the *subject*, rather than the individual.

The concept of the subject is one which has emerged in post-structuralist studies of power, language and ideology, but has recently been incorporated into accounts of on-line identity. Sherry Turkle (1995) complains that, throughout the 1970s and 1980s, the idea that the self is decentred and made up of multiple identities was a marginal one – 'a time when it was hard to accept any challenge to the idea of an autonomous ego' (15). Turkle suggests that 'the normal requirements of everyday life exert strong pressure on people to take responsibility for their actions and to see themselves as intentional and unitary actors' (15). But, and with some cause for celebration, the Internet has changed this situation.

Turkle argues that the Internet 'is [an] element of computer culture that has contributed to thinking about identity as multiplicity. On it people are able to build a self by cycling through many selves' (178) – so much so that '[n]ow, in postmodern times, multiple identities are no longer so much at the margins of things. Many more people can experience identity as a set of roles which can be mixed and matched, whose diverse demands need to be negotiated' (180). In this account of identity-as-subject, Turkle overcomes the widespread essentialist views of on-line identity as some kind of deceptive or fictive representation of another level of a 'real' identity.[10]

Subject theory is a model of individuality which shares some similarities with 'role theory' in that individuals are capable of taking on many different kinds of roles. However, it departs from role theory in proposing that there is no subjectivity outside of these roles. In medium theory, therefore, comparing the behaviour of an Internet avatar with a 'corresponding' off-line identity, does not make sense. Conversely, an off-line subject does not simply 'use' a medium to further communication objectives.

Because a medium already presupposes subjects who are its conduits just as much as the technological medium (print, wires, electromagnetic waves) is, medium theory does not speak of 'users' of a medium. Indeed McLuhan abandoned the reference to media participants as 'users' in his later writings.

From the point of view of the medium itself, to seek to understand the avatar's behaviour by establishing a link between that avatar and an off-line identity will tell us very little compared to understanding the way identity is formed within the medium itself.

Demonstrating the link between different kinds of subjectivities and different mediums helps clarify the nature of the mediums themselves. We can recall McLuhan's claim that the technical aspects of mediums produced certain personality types: bookish, tribal, etc.

If we apply this idea to broadcast and network forms as mediums, it produces valuable insights into the modern dynamics of communication. The media audience is well known as a mass (mass media). In both cases the individual is constituted by the medium itself. The Internet avatar is, *by definition*, constituted in a contextless space from which the very sense of knowledge and power is internal to the medium. However, it is also possible to use the Net as a mirror site for relationships which simultaneously exist off-line.

Medium theory insists on the need to look first at the architecture of each medium, to assess the 'subject position' of actors within that medium. For example, within broadcast, the viewer, listener or reader is constituted in a 'passive' role as message creator, but a very active one in terms of the need to interpret messages. In horizontal 'network' forms of communication, the avatar is constituted as a much more active kind of subject position with respect to the medium. However, where it seems that individuals have an active role in regard to a medium, the dependence and attachment they have to that medium are often disguised. This attachment can sometimes be revealed when flows of communication are interrupted. When a connection is broken, the resulting anxiety can be a measure of the attachment, but also the way in which individuals relate to CITs which allow for multiple connection, such as call-waiting on the telephone, multiple applications running at once on a computer, and channel-hopping on the TV, can be a measure of the power of medium. The greater the urge to check an incoming call or email, to roam channels, websites and stations, and the more fragmented are the communication events, the more the individual becomes a subject enveloped by the medium. When the linearity of communication events is removed, the medium becomes more visible.

Over the past thirty years, the fact that mediums are about practice and form, rather than content, has been the subject of numerous attempted theorizations. These include glance theory (see Ellis, 1982), liveness theory, audience theory and medium theory itself. However, when it comes to understanding an individual's relation to a medium, individual identity often becomes obscured or one-dimensional – an effect of the medium rather than an agent of it. The individual is no longer seen as an autonomous monad with an experience of the real, but as nomadic, fleeting and contingent. However, whilst the individual is no longer positivized, there is a tendency for the impact of mediums to be exaggerated, which is a charge often levelled at McLuhan, Baudrillard and the Krokers.

Christopher Horrocks (2001) claims that one of the shifts in McLuhan's thought is, in the early years, seeing media participants as communicating *through* mediums, and, in the later writings, seeing them as *being* the subject *of* the medium (57). By the time McLuhan first begins to discuss the computer, he abandons the early discourse of viewers, listeners and audiences in which 'the medium is the message' to a later discourse about media *users*: 'in all media the user is the content' (58). However, this quote from McLuhan (in McLuhan and Zingrone, 1997: 280–1) is closer to Baudrillard's understanding of the mass as the conduit of media than it is to user perspectives. But Horrocks interprets McLuhan's shift as accommodating a user perspective suited to wideband Internet: 'With Virtuality, in its widest sense, the use of e-mail, e-conferencing and other tools demonstrates the shift from McLuhan's definition of the user as participant *through* a medium to manipulator *of* that medium' (57). Horrocks argues that the telephone could not be manipulated as a medium, however the personal computer is more 'plastic in its interactive possibilities: A typical personal computer might have e-mail, Web chat, RealAudio, word-processing and a news ticker functioning at the same time' (57) where users may vary their attitude in how they construct their own media environment.

There are two problems here. Firstly, the sense in which McLuhan suggests that users become the content does not permit the agency of manipulators or of agents who have an attitude to the medium in which they are immersed. Secondly, as Turkle and others have well shown, identity in virtual space is infinitely substitutable, and can itself be manipulated.

Reciprocity without interaction – broadcast

We have seen, with Thompson, the description of broadcast as a form of mediated quasi-interaction. The concept of quasi-interaction Thompson adapts from Horton and Wohl's concept of 'para-social interaction' (Horton and Wohl, 1956). In an article written in the mid-1950s, when the meta-psychology of media consumption had barely been analysed, Horton and Wohl explore the idea of intimacy at a distance, and the way in which audiences identify with performers as face-to-face events. 'The new mass media are obviously distinguished by their ability to confront a member of the audience with an apparently intimate, face-to-face association with a performer' (228). This intimacy, however, is not necessarily governed by a 'sense of obligation, effort, or responsibility on the part of the spectator, and indeed lies in a lack of effective reciprocity. … The interaction, characteristically, is one-sided, nondialectical, controlled by the performer, and not susceptible of mutual development' (215). Similarly, in Thompson's typology, the mediated quasi-interaction of books, newspapers, radio and television does not have 'the degree of reciprocity' or 'personal specificity of other forms of interaction' (Thompson, 1995: 84). Nevertheless, it establishes a structured 'social situation' of symbolic exchange in which audiences are able to 'form bonds of friendship, affection or loyalty'. In other words, because it is monological, it lacks mutual interaction but enables strong forms of identification which carry a form of reciprocity.

This latter function of mass broadcast media has been emphasized by the systems theorist Niklas Luhmann. In Luhmann, all forms of communication contribute to the construction of reality, but mass media are peculiar in producing a continual 'self-description of society and its cognitive world horizons' (Luhmann, 2000: 103).

As with Thompson, mass media are characterized by widespread dissemination, and 'anonymous and thus unpredictable uptake' (Luhmann, 2000: 103). This leads to a paradox: 'the reproduction of non-transparency in transparency'. Mass media are self-referential; they may not actually reflect a reality outside themselves. However, what is transparent is that they operate as a reserve for processing information. The individual consumption of media may be unpredictable and anonymous, but as a system, everyone can relate to it with ease.

Whereas mass media provide definite forms of solidarity, however, they do so with little or no interaction. Thompson describes broadcast media as 'mediated quasi-interaction' because they are only semi-interactive. In Thompson's view, only direct interaction, be it face-to-face or electronic ('real-time'), can qualify as a form of reciprocity. Clearly, Thompson feels obliged to include broadcast as a form of interaction, but does not specify why this term needs to be retained when no reciprocity is alleged to be occurring.

What becomes the distinguishing feature of quasi-interaction is not its 'interactivity' in a conventional sense, but that it is able to act as a mechanism of communicative solidarity for social actors. Identification and recognition make up for the lack of reciprocity in such a situation. In this, Thompson redresses the way second media age theorists dismiss or overlook the fact that broadcast, even if it is unequal at the level of *interaction*, still facilitates a powerful form of social *integration* – one which even colonizes electronic interactivity itself – in, for example, the way in which Internet participants aspire to broadcasting their own personal web-page. Thompson also overcomes the unnecessary historicization of media ages that is widespread in contemporary literature.

In order to restore the social aspects of the different types of interaction specified, Thompson also proposes that mediated interaction and quasi-interaction have their own systems of social organization. He conceives of these systems as being built out of the basic building blocks of 'local contexts', which he calls 'interactive frameworks of production'. However, for mediated quasi-interaction, only the producers of symbolic forms are immersed in a 'local' interactive framework of production, whereas for the recipients, who are indeterminate, another framework is added, 'interactive frameworks of reception'.

In the social organization of face-to-face interaction, there is no separation between the context of production and of reception.

In the social organization of mediated interaction, the context of production and that of reception are separated, but the settings of the production of messages are also available for face-to-face kinds of interaction. We may be writing a letter or an email and be able to engage in embodied conversation.

The social organization of mediated quasi-interaction is also a setting for face-to-face and mediated interaction. Whilst watching television we can be on the phone or engaged in conversation with others who are also watching the television. This confluence of media events becomes a form of 'social organization' when '[t]he conversational content of the face-to-face interaction may be determined largely by the activity of reception, as when individuals are involved in commenting on the messages or images received' (Thompson, 1995: 89–90). Thus, many face-to-face conversations have a reference to TV as their content.

However, in the above models of intersecting forms of communication, face-to-face interaction continues to be privileged. Thompson's primary aim is to show the complex ways in which extended forms of interaction mediate the face-to-face.

Thompson's model does not view the extended forms of interaction as themselves social ontologies except insofar as they mediate all of those 'local contexts' of production. Moreover, it can be argued that an oversight of Thompson's analysis is that the very idea of 'local' contexts of communication is made possible by communication that is stretched across space and time. The very experience of the local *qua* local, in its contemporary form, is made possible by more universal and extended forms of symbolic exchange. To appreciate the fact that extended contexts should be regarded just as important as 'local' contexts would also address the problem of describing quasi-interaction as a form of interaction.

Remember that the second media age theorists see the problem of broadcast as being one-way, whereas networked, horizontal forms are two-way. But this holds true only from a logocentric conception of interaction. With the aid of Thompson, we can reveal a sense in which broadcast contains forms of reciprocity.

To return to Figure 3.1 (p. 53) regarding the architecture of broadcast, we can note that at the level of interaction it is a largely one-way form of communication. Thompson notes, in relation to television, that there can be various kinds of 'right of reply' and interviews with audience members,[11] but the overwhelming volume of 'transmitted' messages is one-way.

However, there is another way to view this architecture which reinstates its interactive quality. To appreciate this view, we must abandon the idea that interaction must take place directly between agents, and view it instead as something that can occur within a circuit of symbolic exchange which may involve both human and technical agents.

In the architecture of broadcast that is illustrated in Figure 5.1, the arrows have been located in both directions to suggest a dialogical relationship. The idea that embedded within broadcast is a form of reciprocity is not easy to see. It usually means establishing that the one-way communication that is said to exist within broadcast is actually two-way after all. This means establishing how the audience is a producer of a media form as much as a destination for it.

Broadcast is a communication architecture in which a number of forms of reciprocity are embodied.

- Broadcast isn't entirely monological. There is the reciprocity which occurs when media consumers also become producers. Thompson gives examples with television, but it needs to be appreciated that it is common to all forms of broadcast: 'letters' to the editor, opinion pages, talkback radio and street interviews; and also on television, non-acting

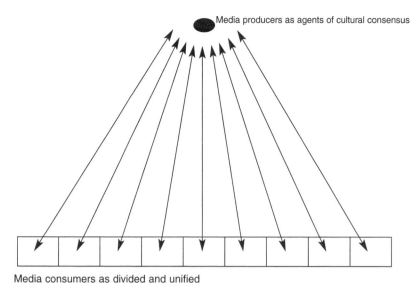

Figure 5.1   Ritual model: high integration/high reciprocity

persons appearing on documentaries, reality TV programmes, and the high visibility of members of a 'live' audience.

- Broadcast is indirectly dialogical via the ratings system. Media producers may shape audience tastes, but, equally, audiences react to trends in broadcast as a whole and may withdraw their patronage, leading to changes in programming.
- At an individual level, media consumers do not simply comprise 'an indefinite range of recipients'. Audiences are specific to definite genres and times, and constitute a remarkably high degree of solidarity. This solidarity is channelled totemically and ritually through 'media agents' – the characters, the presenters, the hosts and the media workers who facilitate the structural architecture of broadcast. It is through these agents that individual members of a given audience indirectly 'interact' with each other.[12] Instead of having a directly horizontal communicative relationship with others, a detour is taken via these media agents. It enables a form of the many speaking to the many via the performative quality of the apparatus proving itself in every act of broadcast.[13]
- As Thompson points out, the peculiar form of quasi-interaction consummates itself directly between audience members when they find themselves in a face-to-face interaction. In fact the reason they may associate is because of the common bond they feel with media agents who have already brought them together. They may otherwise routinely associate face-to-face, but make of broadcast media a primary basis for mutual conversation.

However, to continue to view this architecture in terms of 'interaction' becomes problematic, a matter which even Thompson has reservations about, in his ad hoc remark that what he calls quasi-interaction should really be called quasi-participation, which nevertheless sustains reciprocity as per the above. What is missing from Thompson's account is the way these agents of media (the media workers, the culture industry) carry a form of social integration, what James and Carkeek (1997) have called 'agency-extended' integration, which we shall return to below in the section on 'levels of integration'.

It is insofar as broadcast can be regarded as a form of reciprocity or 'quasi-social' or 'para-social' interaction that the second media age thesis becomes unsustainable. Broadcast can only be considered a one-way form of communication to the extent that the metaphor of the media as channel, rather than environment, is adhered to. Broadcast media enable a form of reciprocity without interaction in which many individuals are 'metaphorically' interacting with each other constantly.[14] The broadcast medium becomes the agent through which each audience member is able to 'reflexively monitor' what it is that other audience members are consuming. Of course, if a broadcast programme is consumed in the mutual presence of others, this reflexive monitoring will bring in their reactions.

In Figure 5.1, therefore, broadcast, like network forms of interactivity, is characterized as a form of the many speaking to the many. This can be appreciated only if broadcast is viewed as a medium of social integration. The main relationship that is active for the audience is with other audience members, not with performers and celebrities. The latter are merely the conduit by which solidarity is achieved with other viewers, listeners and readers. Here, we can take issue with Meyrowitz's *No Sense of Place* (1985) for characterizing para-social interaction as illusory. Certainly individual members of audiences may come to feel they '"know" the people they "meet" on television in the same way as they know their friends and associates' (119), but the intimacy being established is really with other members of the audience, most of whom they will never meet.

Broadcast, like network activity, when conceived either as a technical environment or as a form of social connection, is able to facilitate a sense of belonging, security and community, even if individuals are not actually directly interacting. Both these mediums, in different ways, enable forms of social integration rather than extensions of face-to-face interaction. They offer modes of relating which can determine the form of general interaction in a given media society (see Table 5.3).

Broadcast integration also brings about a high level of recognition between audiences and media producers, but, as we have seen, low levels of actual interaction. In such a mode of integration, audiences come to identify strongly with media presenters, news teams, and film and soap stars (see especially Langer, 1997). Some actors can readily acquire a cult status whilst many news programme presenters may be endowed with authority. All are bestowed with charisma as a reflex of the concentration

Table 5.3 Broadcast and network as forms of communicative integration

Broadcast integration	Network integration
The many interact with the many by way of the agent of message procedures ('media workers', the culture industry, etc.)	The many interact with the many by way of computer simulations of presence
High level of recognition/identification	Low level of recognition/identification
Very low level of interaction	Very high level of interaction
Individual experiences strong identity/ identification with figures of authority, charisma or cult movements	Individual experiences weak identification with others as figures of authority or charisma
Concentration spans of audiences are sold to advertisers	The need to communicate in highly urbanized settings is sold to individuals

Source: from Holmes, 1997: 31.

of consciousnesses in their person. The intensity of such concentration can have substantial consequences for the distribution of recognition relations within the specular field of a given broadcast medium. As discussed in the next chapter, celebrities may become over-exposed, whilst audiences can be too dispersed for cult fixation to gather momentum. These kinds of changes will have an effect on the kind of virtual community that is constituted by media events as much as entire mediums.

But there is also the question of the number of broadcast channels that are available in a given national frame. The 'nationwide' audience which Morley (1980) first theorized can only occur in settings where there are a relatively small number of channels and broadcast conduits. In Australia, commercial networks have recently staunchly opposed the introduction of multi-channelling, 'which would reduce their capacity to offer advertisers a mass market on a single channel, while bringing in no extra revenue' (Tingle, 2002: 1). In order for advertisers to have an effective mass audience, the number of broadcasters either has to be very small or very specialized, as in the case of cable TV.

Interaction without reciprocity – the Internet

To turn now to the second column of Table 5.3, just as there can be reciprocity without interaction, there can also be interaction without reciprocity. This is exemplified by any form of interaction between strangers at a distance, as in much of the communication on the Internet. Such forms of interaction are possible in fleeting, transient, face-to-face contexts in abstract settings like a large metropolis where there is little likelihood of such communication being repeated. But on the Internet it becomes a systemic reality. Most Internet identities are avatars for whom reciprocity is not possible. For reciprocity to be successful there has to be some sense of obligation to communicate which has a socially organized basis. Except for mirror sites on the Web, and Internet email use, none of the sub-media of the Internet provide a stable environment for relations of trust to develop between users, or between users and information. For example, on the World Wide Web, the phenomenon of bit-rot has even deterred authors and analysts of cyberspace from print-publishing URLS from websites (see Lunenfeld, 1999: 237). In CMC Net sub-media, avatars are not accountable or responsible to each other except insofar as they apply pre-given norms from the off-line world. But to do this is precisely contrary to what an avatar is.

To appreciate this we need to distinguish between communication on the Internet between interlocutors who have a prior face-to-face or institutional association and those who are anonymous to each other. In the former circumstance, Internet communication is largely an affair of the conduit and vessel. It may be simply a more efficient or instrumental means of sending messages to those who are already known.

However, a large part of Internet use is between individuals who do not know each other from other contexts, even institutional contexts. A great deal of interaction takes place mediated by a computer-generated infrastructure, but the obligation of reciprocity does not exist. Reciprocity requires an identification of interactants in order for an exchange to be rendered mutually. The proof of the distinctive consequences of anonymous communication on the Internet can be seen in the fact that it is considered necessary to have 'policies' to deal with such problems (Kling et al., 2000). Such policies are not required in broadcast mediums. As Kling et al. argue: 'While many people believe that anonymous communication on the Internet is not only acceptable but has positive value, others see risk in it because anonymous users are not accountable for their behaviour. Consequently, anonymity can mask or even encourage criminal or anti-social behaviour'[15] (98). To remedy this perceived problem, the American Association for the Advancement of Science's (AAAS) Program in Scientific Freedom, Responsibility and Law held a conference in 1997 to 'better understand the nuances of anonymous communication on the internet and develop ideas that could guide policy development in this area' (Kling et al., 2000: 98). Four major principles were advanced as providing a guiding role in policy development:

- Anonymous communication on-line is morally neutral.
- Anonymous communication should be regarded as a strong human right; in the United States it is also a constitutional right.
- On-line communities should be allowed to set their own policies regarding the use of anonymous communication.
- Individuals should be informed about the extent to which their identity is disclosed on-line (Kling et al., 2000: 99–101).

As demonstrated by the AAAS exercise, the idea that the 'rights' of avatars need to be protected is again a function of their low visibility, and the fact that no kind of 'other', be it a person or an authority, has a commitment to an avatar. Very weak identification with others is experienced between Internet avatars. There is a very low level of recognition between them as they lack 'off-line' contexts of recognition which can provide wider bases of identification. Often recognition is limited to text-based interaction, in which the identity of an other is confined to what he or she can construct with text.

However, it need not be the case that Internet avatars are deprived of reciprocity. They could associate at another level of interaction, by being members of the same institution, or by revealing a social world in which they could relocate their identities, their character, and their reputation; or, finally, they could meet each other. Any of these other contexts would enable a triangulation of each interactant's identity. Nonetheless, the sheer volume of traffic within sub-media of the Internet ensures that most interaction is between avatars.

In cases where, from the point of view of a given interactant, interaction is entirely internal to CMC, the kind of identity which is formed may be described as constituted at a merely 'intellectual' level of abstraction which promotes a certain kind of solipsistic ego formation. Insofar as an interactant can discontinue a relationship at any time without its having repercussions in any kind of socially constituted field of recognition, CMC is well suited to the generalization of an autonomous individualism which has long been characteristic of intellectual culture. Instrumental control and liberation from the flesh is at the core of such individualism (see Sharp, 1985). Interactants who do not like the responses they get from interlocutors do not have to confront them at all. They can control the kinds of interaction they have by minimizing random contact and only continue those relationships in which their ideational reflection shines the brightest. Ultimately, such means of control result in avatars having conversations with no one but themselves, particularly given that their identity is a self-contructed-for-others which engages with a myriad of other 'selves-constructed-for-others'. This is not to say that such selves are not 'real'; on the contrary, they are constituted 'cybernetically', as it were, and willingly participate in a system which mutually reinforces the maximization of each interlocutor's own reflection.

The levels of integration argument

Thompson (1995) points out that for most of human history communication has been face-to-face. Most human institutions have evolved within the scope of face-to-face relations. The emergence of new types of interaction which Thompson specifies has brought about new social fields and changes to the nature of work, public/private divisions and leisure time.

But it has, most significantly, brought about new modes of *social integration*. In order to explain this we need to formalize the distinction between interaction and integration. Whereas interaction involves the empirical act of engaging in a speech act, either extended or in mutual presence, social integration is made possible by some or other form of reciprocity, via interdependence, long-term continuity of association and strong identification with an other – even an abstract other.

The claim here is that reciprocity can still occur without direct *interaction*. In fact it is possible to see how *most reciprocity involves little direct interaction at all*, but rather is embedded in numerous kinds of rituals which solidarize certain kinds of communion which may not be empirically obvious.[16] From a sociological point of view, these rituals need not involve restoring co-presence at all; rather, they may be oriented towards quite abstract forms of association – but association nevertheless.

Social integration through ritual

A recent important book which can help us see how reciprocity in media sociality can and does occur without any form of interaction is Nick Couldry's *Media Rituals: A Critical Approach* (2003). I am in complete agreement with Couldry's claim that '[w]e cannot analyse the social impacts of contemporary media without taking a position on broader social theory' (3). And Couldry takes as the indispensable starting point for such a position a radicalized reassessment of Durkheimian thought on social integration.

As Durkheim allows us to ask: 'how, if at all, do societies cohere, how is it that they are experienced by their members *as* societies?' (Couldry, 2003: 6).[17] Couldry follows Durkheim in suggesting that even modern, complex media societies exhibit principles of coherence that can be understood through the study of rituals. Three senses of ritual are distinguished: ritual as habitual action, ritual as formalized action, and ritual as action that is associated with transcendental values. It is this latter form of ritual which has the most bearing on the role of media in social integration. For Couldry, media represent a 'wider space of ritualization' which has a range of transcendental functions beyond individual habits of media consumption. Moreover, the portrayal of 'already existing ritual action' such as the televising of a religious event fails completely to illustrate what he wants to capture in the notion of media ritual.[18]

For Couldry, media rituals occur within psychological and spatial boundaries in which it is possible for strangers to interact in symbolically unequal yet naturalized ways in which they experience a shared set of values. For example, in a description similar to Horton and Wohl's 'para-social' interaction, Couldry discusses the ritualized meeting between the 'media person' (celebrity) and the 'ordinary person' (media consumer) in which audiences live their world of values through the celebrity at the same time as the celebrity somehow comes to stand in for, and provide a pivotal window into, social life (26–7).

This concentration of symbolic power is, for Couldry, more pronounced in the modern media than in any other social institution (like church and education). He favours '[a] *strong* concept of symbolic power ... [which] ... would insist that some concentrations of symbolic power (for example, the concentration from which contemporary media institutions benefit) are so great that they dominate the whole social landscape' (39).

Couldry sees media as supplying a late-modern 'myth of the centre'.[19] He describes a condition in which every member of modern societies believes that the media are our 'access point to society's centre'. This is, of course mythical, yet, to the extent that it is universally accepted, also very powerful. The media themselves also enjoy a virtual monopoly over the power of naming (43),[20] and 'defining-the-situation'. However, this power is not a kind of hegemony in the Gramscian sense, or even the kind of 'will to truth' which Foucault formulates; instead it is much more a relation to the media's status as an institution in the neo-Durkheimian sense, rather than in the political spirit.

> Because society's symbolic resources are very unequally distributed (with media institutions being the main beneficiaries of that inequality), these ongoing conflicts of definition are marked by symbolic violence: certain def-initions have enough weight and authority to *close off* most other alterna-tives from view, although such closure can never be total and is always, in principle, open to challenge. (43)

The rest of Couldry's book looks at genres of media where this authority can stamp itself the heaviest: the 'liveness' of the media event, the 'reality' TV show, and the way in which fan clubs, in defining themselves purely in terms of media, ritually reproduce the 'myth of the centre'.

The distinction between 'levels' of integration

Couldry's book is very useful for affirming and renewing the significance of the concept of social integration and how it is related to ritual in media societies. However, in a given society, 'social integration' need not be seen to occur in a singular homogeneous space that changes over time.[21] Rather, we can also point to perspectives which propose that individuals can source their sense of integration from a range of levels of association, which primarily differ in their degree of 'abstraction' in space and time, from embodied forms of intimacy to the generalization of 'action at a dis-tance' which characterizes contemporary global culture.

The particular social integration perspective which is to be introduced here has a number of exponents, who, from varying and quite separate moments in sociological theory, can all be written into the development of a distinctive tradition. This framework can be vicariously observed in a number of different intellectual movements: situational/interactionist (Goffman; Thompson/Meyrowitz), phenomenological (C.H. Cooley/ G.H. Mead; Calhoun), and abstraction arguments (Durkheim/Sohn-Rethel; Giddens/Sharp/Slevin).

Situational/interactionist perspectives

The work of Joshua Meyrowitz provides an invaluable resource for problematizing the significance of New Media. Meyrowitz's writings are distinctive in the way they synthesize the media writings of McLuhan and the sociological work of Erving Goffman. In his pre-Internet work *No Sense of Place*, Meyrowitz (1985) was beginning to integrate communication theory and sociological accounts of everyday life in useful ways.

Meyrowitz's use of Goffman is in taking up social role theory to explain different levels of association. McLuhan's notion of sense ratios is abandoned in favour of 'face-to-face', 'back region' and 'front region'. In this view, there are many kinds of selves distributed across levels of public and private 'regions'.

In adopting this approach, Meyrowitz (1985) explores 'a common denominator that links the study of face-to-face interactions with the study of media: the structure of "social situations"' (4). The key premise from which he conducts the analysis is that media have architectures which shape social situations in profound ways. Media create environments and various forms of electronic assembly which can either cut through spatial segregation or replace it with electronic versions.

Focusing mainly on the case of television broadcasting, Meyrowitz primarily wants to show how electronic media can break down the 'traditional association between physical setting and social situation' and teleport individuals into an electronic public sphere (7).

> Imagine that many of the walls that separate rooms, offices, and houses in our society were moved or removed and that many once distinct situations were suddenly combined. Under such situations, the distinctions between our private and public selves and between the different selves we project in different situations might not entirely disappear, but they would certainly change. We might still manage to act differently with different people, but our ability to segregate encounters would be greatly diminished. We could not play very different roles in different situations because the clear spatial segregation of situations would no longer exist. (6)

All of the 'backstage' behaviours that are carried on in the cloistered architecture of interaction would now be visible by larger audiences with better visibility. 'Media, like walls and windows, can hide and they can reveal. Media can create a sense of sharing and belonging or a feeling of exclusion and isolation. Media can reinforce a "them versus us" feeling or they can undermine it' (7).

Importantly, however, Meyrowitz resists the temptation, commonplace in much Internet literature, to see New Media as having abolished face-to-face interaction. Rather, for him, New Media expand the range of possibilities of how individuals might interact, which can sometimes present conflicting senses of the normative context in which actors are able to take on roles. This may include confusion about which medium to undertake interaction in, and the nature of expectations individuals feel about having to speak, to reply or to listen. Thus the title of Meyrowitz's book *No Sense of Place* alludes not to the disappearance of a sense of place, but to the saturation of the self by a clash of ontological 'levels' of association.

In a communicative culture dominated by television, individuals are typically able to hide behind the medium of broadcast when they are actually interacting with a television itself. Without the comfort of the television on, even in the background, a kind of role-vacuum is created, in which responsibility shifts to the individual to be self-active.

Meyrowitz's analysis is well complemented in Thompson, but with many of the conceptual shortcomings we identified above. Nevertheless, Thompson also makes a thoroughgoing case for a 'levels' argument, even though it does not carry a sense of integration. He claims that entire systems of social organization are based on these levels, but it is as though such organization is functional to dynamics of techno-social systems rather than fulfilling different kinds of needs that emerge out of changes in the infrastructures of communication.

Calhoun's phenomenological levels of socialization

Working from what can broadly be described as a phenomenological approach, the American public sphere theorist Craig Calhoun has made some progress with developing a levels approach which is an advance on those of Meyrowitz and Thompson. In three important articles on computer-mediated social relations, Calhoun (1986, 1992, 1998) innovatively develops the idea of indirect social relationships. Following C.H. Cooley's work in *Social Organization* (1909), Calhoun works up a typology-driven model of communicative levels of social integration. Where Calhoun differs from Thompson and Meyrowitz is in placing social *integration* rather than interaction as the traversing agency across these levels. To explain this we need to revisit Cooley for a moment. In *Social Organization*, Cooley proposes the need to distinguish between primary and secondary social relationships. 'A primary relationship must be both directly interpersonal and involve the whole person.' A secondary relationship, by contrast, 'need meet only the criteria of directness', but not in a way which permits any kind of intimacy or many-sided recognition (Calhoun, 1986: 332).

Calhoun wishes to add tertiary and quaternary 'indirect' relationships: 'Noting the impacts of modern communications technology, we may go further and identify as indirect those relationships that require the mediation of a complex communications system' (332).[22]

For Calhoun, tertiary relationships are ones that individuals are 'aware of' and active in, for which he lists bureaucracy as an archetypal form. 'We have "tertiary" relationships with those to whom we write to complain about errors in our bank statements, with our political representatives (most of the time), and, often, with the senior managers of the companies for which we work' (332). Quaternary relationships are ones which we are not aware of such as surveillance infrastructures, and we are exposed to techno-social systems in which we find ourselves unwilling participants (333).[23]

Both tertiary and quaternary relationships allow for what Calhoun calls large-scale social integration, the definitive locus of which is the modern 'mega-urban' city. Cooley's secondary direct, but unfulfilling, relationships are, in some measure, a part of the large-scale urban picture because they offer 'serendipitous contact across socio-cultural boundaries' (335).

But secondary relationships are also cause for the experience of widespread anomie, precisely because of their practical difference from primary relationships. Calhoun argues that this difference is ontological, not simply a matter of perception. Secondary relationships are generally held in low esteem by city dwellers, as advanced by Cooley himself at the beginning of the twentieth century. Primary relationships, found in family and face-to-face networks, provide spontaneous settings of integration even when they involve conflict.[24]

The frustration of secondary relationships, in workplaces, in the market place, in the public sphere, is that they take up so much of our time, and are emotionally involving but unfulfilling. Whilst it is true that primary relationships may also be unsatisfying, at least they are capable of generating enduring loyalty and satisfaction, which secondary ones can't. Secondary relationships foster a destructive notion of freedom in which 'strangers often seem to exist *only* to annoy us' (as Sartre once suggested, 'Hell … is other people'), and such 'relationships are simply the choices of the moment rather than commitments' (335, my insertion). They are purely functional, such that when even their functionality fails, it reverberates as an even more intense condemnation of the hopelessness of the emotional or other value of such levels of association.

Under such conditions we seek to avoid emotional involvement in our dealings with strangers and 'deal with problems by trying to escape', as narrated in Philip Slater's (1971) account of the 'pursuit of loneliness'. Such a condition has also become the subject of films like *Falling Down*.

The ontological impasse between primary and secondary relationships, which is in some sense 'proven' by the everyday tension between them, is, argues Calhoun, eased by the widespread development of tertiary relationships. Whilst some may see technologically mediated relationships as just a disembodied extension of estranged secondary relations[25] (particularly when tertiary relationships are only a rudimentary or modest feature of social relations generally), for the most part, he argues, such a level of relationship can be experienced as emancipatory. Remember that Calhoun was advancing this thesis well before the utopian discourses which heralded the Internet as relieving everyone from the impersonal aspects of trying to maintain large-scale integration in an embodied form by way of networks of agents.

The 'proliferation of tertiary relationships cuts down on secondary, but not primary, relationships' (336). Calhoun argues that in substituting for the unwieldiness of large-scale social integration occurring at an embodied level, a tertiary relationship can actually free up individuals to spend more time in primary modes. 'We might focus time and energy on community building, friendships and family life, though this is only a possibility, not an automatic result' (336).

For Calhoun, this possibility is a feature of all technologically extended and mediated relationships, not simply communicative ones. He gives the example of the automatic teller machine. 'Direct interpersonal contact is reduced, as the customer no longer deals with a teller. But the customer also spends less time standing in lines and has greater flexibility as to when to use banking services.' The customer does not have to endure the

> rebuff of non-recognition. ... There is often a disappointment on the customer's side at not being recognized (and apparently not trusted) by a person with whom he or she may interact on a regular basis. ... It is not obvious that we are losing much of value in giving up this sort of 'personal' interaction. (336)

Conversely, argues Calhoun, the flexibility we have with interfacing with the much more numerous machines frees up time which can be used more productively elsewhere, as well as being 'redeployed into primary relationships' (336).

However, Calhoun's caveat is that while computers might greatly assist in large-scale integration,

> there is as much (or more) reason to think that computerization and new communications technologies will lead to or accompany further deterioration of interpersonal relationships. A drift toward relationships of convenience might be accelerated; passive enjoyments from the mass media might predominate over active social participation. A few people might even wind up preferring relationships based on single common interests and mediated through computer networks – or worse (from the point of view of social integration), preferring the company of computers themselves, which are dependable, don't talk back, and don't make silly mistakes (very often). (337)

In other words, Calhoun perceives a tension between the capacity of tertiary relationships to enhance and regenerate primary ones and their tendency to replace them altogether.

Constitutive abstraction

A position known as 'constitutive abstraction' has been advanced for many years by an Australian group associated with the journal *Arena* who have developed a social theory which looks at social relations as an intersection of 'levels of integration'. It is an account which has emerged largely in isolation from other approaches to 'levels' theory but is one which nevertheless contains invaluable points of innovation.

The *Arena* thesis takes up the work of Alfred Sohn-Rethel, who poses the question in his *Intellectual and Manual Labour* (1979): can there be abstraction other than by thought? By this, Sohn-Rethel suggests that abstraction is not merely a property of the mind, but can occur in social relationships also. When it does so, these relationships are no less 'real' than less abstracted ones; however, they do possess distinctive features.

The constitutive abstraction argument shares the departure point of Thompson that most human culture has been framed by face-to-face relationships and only recently have we seen the emergence of new forms of social relations. The newer forms are shaped by the rise of postmodern technoscience of all kinds, not just communication. Postmodern technoscience differs from the science of the Enlightenment or modernity in that it is seen to reconstitute the natural and social worlds rather than simply harness such worlds in the name of progress. Thus, IVF fundamentally intervenes in the process of human reproduction, to the point of making many of its qualities redundant. Nuclear power, in sub-atomically rivalling processes which occur on the sun, makes possible the 24-hour society, in a different way than did coal.

The main agents of these changes are the intellectually related groupings. The results of their practice of reconstituting the world creates settings of 'new nature' which abstract all persons into an orbit of interchange and exchange that can be seen to be derived from intellectuals and their practices.

In this view, therefore, print, the dominant medium of intellectuals, is seen as a precursor to modern technologically extended forms of communication. The newer forms of communication bring about practices which abstract the individual from more embodied forms of social relations and de-link those individuals from the kinds of roles that they once had, giving them the appearance of autonomy.

As Sharp (1985) suggests:

> The first point to make about this sort of practice is that in order to be engaged in it the person must be abstracted from the settings which make

up the structure of what we still take to be the mainstream life of society. Of course such members of the intellectually related groupings go to work, get paid, have families and go to the supermarket and the pub; but none of these settings accounts for or defines the specificity of the relations within which they carry on their distinctive practice. The second point relates to the social form of which this abstracted practice is constitutive. Basically it would seem that for the intellectuals we can say that interchange is mediated by print which serves as one abstracted way of symbolizing the linguistic element of face-to-face interaction. This technological medium allows the social tie to be extended in space and in time. It creates a setting whereby the participant is 'lifted out' of the relationships of everyday life and where at least subjectively persons experience themselves as the authors of their own creations. In other words, they begin to experience themselves as post-individual or, as the contemporary pop term would have it, as 'autonomous persons' acting in a setting where the boundaries or constraints visible in ordinary life might seem to have dropped away. (62–3)

However, the emergence of the autonomous individual who no longer has the roles which once seemed to be easily ascribed is not merely a matter of changes in interaction. For social relations generally to take on the mode of interchange which has characterized intellectuals for many centuries,

different modes of the extension of the social relations must emerge – in transport, in communication generally. To illustrate this process for the polity, the extended process of commodity exchange and the reconstruction of the practices of ideological integration, it is scarcely necessary to look beyond the role played by television as a form of extended social relationship. But the ways in which the population at large is drawn within the field of extended interaction all require separate treatments in their own right. (63)

It is these 'separate treatments' which are quite underdeveloped in the *Arena* thesis. However, Paul James and Freya Carkeek (1997) have proposed levels of integration in a way in which New Media can be more readily contextualized. Drawing on the *Arena* framework, they explore 'levels of social integration', 'understood as intersecting forms of structured practices of association between people' (110). The levels do not exist as pure forms but are analytically distinguishable. Nominally, they distinguish three levels: face-to-face integration, agency-extended integration and disembodied integration.

Face-to-face integration is defined as the level where the modalities of being in the 'presence' of others constitute the dominant ontological meaning of interrelations, communications and exchanges, even when the self and the other are not always engaged in immediate face-to-face interaction. Under such forms of interrelation, the absence of a significant other, even through death, does not annul his/her presence to us. Agency-extended integration involves the extension of possibilities of interrelation through persons acting in the capacity of representatives, intermediaries or agents of others. (111)

This level corresponds to what Cooley and Calhoun call 'secondary relationships'.

'Disembodied integration is the level at which the constraints of embodiment, for example being in one place at one time, can be overcome by means of technological extension – broadcasting, networking or tele- phoning, to name only a few' (111). This level corresponds to Thompson's mediated interaction and mediated quasi-interaction.

James and Carkeek go on to clarify that 'each of these levels is more abstract than the level "prior" to it, and each is implicated quite differently in the ways we live the relationship between nature and culture, and the ways we live our bodies and the "presence" of others' (111).

As posited by the *Arena* perspective, levels of integration are not empirical forms, however, but are 'ideal-type' social forms, which are also capable of co-presence within a given *societal* form.[26] In tribal societies it is face-to-face meetings which are the most important, whilst abstraction at the level of religion is very much in terms of the totemization of the body and of the land. In agency-extended society, the greatest importance is attributed to manual craft and networks of actors consolidated in institu- tions. In more abstract, information societies, on the other hand, it is the intellect which is ritualized in the fetish of high-technology and ultimately the use of technologies which presuppose the analytic dismembering or displacement of the natural world.

However, 'less abstract' is not equated with underdeveloped, nor is 'abstract' equated with developed. Most literature is framed by such a linear model of the succession of forms of mediation, but in the *Arena* model, even though new levels of abstraction keep emerging on an historical basis, they do so in tension with established forms, which they intersect with rather than simply supersede. Thus, *Arena* authors claim not to argue for a return to close-knit parochial communities which rest on face-to-face relations but, rather, that the tension between levels is enriching 'as long as any one level does not come to constitute the dominating mode of living- in-the-world, and so does not thin out prior levels of human interrelation' (James and Carkeek, 1997: 111).

For James and Carkeek, these different levels intersect in complex ways. Each level may 'contradict, qualify, dominate or be "thinned out" by other levels of integration' (111). In particular circumstances, one of these levels may become dominant throughout a societal form. However, it is older forms of integration which are vulnerable to annulment by newer forms, rather than the reverse. In tribal societies, face-to-face inte- gration is generally dominant, but seldom comes into tension with agency- extended or disembodied integration. What is of interest in this situation is the way in which a tribal, status-bound culture might take up disem- bodied forms of communication. In this circumstance what is important is that persons are bound to each other even when the self and the other are not engaged in immediate and embodied interaction. Therefore the point of 'finding connection' by means of tele-mediation may seem anachronistic for persons formed in this way.

When agency-extended integration predominates, it is the expectation of embodied networks of intermediates which becomes the dominant centre of 'ontological security' within a societal form. In such a setting, word of mouth, recommendations, networks of knowledge and activity offer far more of a guarantee of bonding to the social form than can be achieved via an abstract market, telecommunications, etc. The dominant form of this level of integration is the institution: church and state, guild and corporation, media and the culture industry – and their various constituencies, which each become trusted anchors.

Like agency-constituted integration, the interactive events of disembodied integration usually refer back to the face-to-face (e.g. emoticons, cybersex) whilst annulling the face-to-face *by extension*. Disembodied integration is the most paradoxical because it typically creates the very conditions which it nostalgically attempts to overcome.

Extension of communication by agent

An important level of communicative integration which the abstraction thesis points to, but which is too often overlooked in current-day accounts of 'community' and 'interaction', is that of extension by agent. For example, almost always, analyses of technologically extended social relations limit themselves to a comparison with 'face-to-face' communication, from which follow familiar binaries of embodied/disembodied, virtual/real, etc.

What is overlooked is the way in which a communication process which also forms part of the 'reciprocity without interaction', which we discussed above, is mediated by other actors, and is not simply a technical means of transmission. This is more visible in broadcast than with the agents who are at work in CMC – the software designers and programmers. In the broadcast situation, Raymond Williams (1961) discusses an important distinction between *source* and *agent*. A *source* is someone who offers an 'opinion, a proposal, a feeling' and 'normally desires that other persons will accept this and act or feel in the ways that he defines' (293), whereas an *agent* is someone whose 'expression is subordinated to an undeclared intention' (293), such as attracting audiences, editing a text to satisfy certain tastes, etc. 'In social terms, the agent will normally in fact be a subordinate – of a government, a commercial firm, a newspaper proprietor' (293), necessary to any complex administration.

> But agency is always dangerous unless its function and intention are not only openly declared but commonly approved and controlled. If this is so, the agent becomes a collective source, and he will observe the standards of such expression if what he is required to transmit is such that he can wholly acknowledge and accept it – recreate it in his own person. (293)

Thus, for Williams, agents have a moral obligation to give carriage to what is commonly approved and controlled, and where the agent unre-flexively assists in the process of transmission, without regard for the standards of expression, he or she is acting in a way that is 'inferior to the poorest kind of source' (293).

To follow Williams' argument, agents are a necessary part of any kind of mass communication, but they are also significant actors in the 'medi-ation' of culture itself. They figure deeply in the way that reception and response depend on factors other than the 'techniques' and means of transmission. But they are so often overlooked as mediators of communi-cation because of the fact that dominant and popular conceptions of the communication process make too sharp a distinction between source and technique, an author-message and conduit (cf. Meyrowitz), not leaving much room in between for the idea of *agent*.

Abstraction and the Internet

A much more recent abstraction argument which provides an account of the Internet from a social theory point of view is that of James Slevin, in his useful volume *The Internet and Society* (2000). Slevin draws together theoretical analyses from Anthony Giddens, John Thompson and a theorist of postmodernity, Zygmunt Bauman.

Giddens prefigures the constitutive abstraction argument discussed above, but in a less developed way. Nevertheless, Giddens' more familiar concept of 'space-time distanciation' is one suited to developing a contex-tual analysis of the Internet.

Giddens argues that traditional forms of institutional socialization have declined as our occupation of social space is today less tied to a sense of physical place (see Giddens, 1990; this is also integral to Thompson, 1995). In Giddens' account it is not merely new communication practices which contribute to such de-physicalization, but also monetary exchange, travel, chronological time-keeping and old communications technologies such as print. The distanciation characteristic of these techno-social arrangements produces a '"lifting out" of social relations from local con-texts of interaction and their restructuring across indefinite spans of time-space' (Giddens, 1990: 21). The 'disembedding' of close-knit social relations results in more abstract forms of social tie, in which time-space relations are no longer experienced as restraints on actors (Giddens, 1987). The for-mer characterize band societies and agricultural communalism, whereas civilizational (pre-industrial) societies and industrialized societies are characterized by successive forms of disembedding (cf. Giddens, 1987: 93–6). Giddens (1987) argues, 'The level of time-space distanciation char-acteristic of band societies is low. The mobile character of the society does not involve a mediated transcendence of space: that is to say, it does not, as in large societies, involve regularized transactions with others who are physically absent' (93–4). For Giddens, these regularized transactions become system-like within time-space envelopes that have their own distinct set of time-space conditions that are tied to technology, institutions and social organization.

James Slevin (2000) adopts the ontological contours of Giddens' account of time-space distanciation by way of its influence on Thompson's theory of communication and cultural transmission. Slevin's particular contribution is to rework the disembedding thesis by arguing that the Internet combines three aspects of cultural transmission in a unique way: as a technical medium, as an institutional apparatus, and as a form of space-time distanciation.[27]

A technical medium (e.g. print, analogue or digital communication) is distinguished by its capacity to store information and reproduce it (as in 'mass reproduction', for example), as well as its availability for participation. The Internet, Slevin argues, is a powerful super-medium which is capable of strongly realizing all of these attributes. However, we cannot understand the significance of this without also understanding institutional contexts which govern the technical medium. For example, it is very easy to eulogize the Internet's capabilities for interaction, but it would be naïve to do so without also pointing out its potential for surveillance. The ways in which the Internet is used lie outside its material/technical substratum, and have more to do with the modern state and its institutions, which have their own culture. Thus, surveillance, for example, could be said to have always been practised by the modern state, but the technical means of carrying it out have changed.

There is yet one other form of change, however, which Slevin adopts from Giddens and Thompson, which is the '*degree of temporal and spatial distancing* involved in the circulation of information and other symbolic content' (Slevin, 2000: 69). Slevin argues that technical and institutional apparatuses of the transmission of culture not only produce time-space relations, but also respond to them. For example, disembedding may facilitate greater numbers of persons who will never meet each other to virtually interact, but it can also contribute to conditions, like globalization, which require more complex and powerful technical and institutional means of connection. 'The pressure and opportunities for mobilizing time-space during the exchange of information constitute the "grounding" for the way in which such exchanges are organized and sustained' (69).

For Slevin, the Internet provides a 'grounding' for a new level of disembeddedness, which he contrasts with mass communication, but in a way in which Thompson's interaction-based comparison is revised. Four key aspects of mass communication are challenged by the Internet as super-medium. Firstly, the Internet is a relatively 'open communication system' which does not require 'large scales of expert systems for the production of content'. Second, the Internet does not just 'equalize' the relation between sender and receiver, but blurs the dichotomy between the two, although this 'may vary from encounter to encounter, from application to application' (74). Thirdly, to the degree that institutional arrangements allow, availability in time-space is enhanced. 'Organizations can now store masses of information on their websites and achieve around-the-clock availability' (75); in comparison, news in mass media will only be broadcast if it is televisable, newsworthy, etc. Finally, traditional mass media are compelled to circulate symbolic forms in ways which parallel Internet communication. Slevin points to the digitalization of broadcast and the proliferation of channels and their specialization in programming and audiences (76).

In each of these cases, the Internet is seen to recast the distinct time-space conditions of electronic media in general. Ultimately, Slevin does not oppose the Internet to broadcast; rather, the two forms must be considered in relation to institutional forces, as comprising a complex level of extended integration that is distinguishable from face-to-face and agency-derived modes.

Conclusion

The different levels arguments that we have explored differ from interaction perspectives in that they suggests that media of communication can act as bases of association which reach well beyond the communication events they make possible. Meyrowitz's understanding of media 'architectures' provides one kind of model for thinking about the hierarchies of structured communication. It is a decisive advance on Thompson's model, which rests on an empirical reduction of 'interaction' (rather than integration) whose basic analogue is the face-to-face; as such, it treats all other interactions as 'mediated'. 'Mediation' theory is a variant of the reproductive view of communication: that mediated or extended communication is a continuation of dialogical interaction by other means. The integration argument insists that these ontological levels are constitutive of distinct orders of the distribution of recognition relations, rather than the mediation of some kind of 'building-block' form of interaction: the face-to-face. In contrast, approaches like Calhoun's and the *Arena* thesis are concerned to show that the division between face-to-face, extended and technically constituted is an ontological one, and that social integration via specific levels is more than simply the aggregation of communicative events within each of these levels.

Unlike interaction theory, which is derived from 'transmission' accounts of communication, and the idea that development in the *means* of communication is about moral improvement, the integration theorists view human culture as being engaged across a range of levels of communication. The sociological basis of the integration perspective resists the tendency of interaction approaches to view the *telos* of communication as providing a transcendental unity, a virtual community or global village.

Rather, what is true of all of the integration perspectives is that in each of the 'levels' of social integration, individuals are separated and united at the same time. It is the architecture of this separation and unity and the tension that exists between levels which will determine the kinds of community and association that are possible within given social formations. It is to the question of community that we shall now turn.

Notes

Some of the passages discussing the work of Derrida in this chapter are derived from Holmes, D. (1989), 'Deconstruction: A Politics Without a Subject', *Arena*, No. 188: 73–116.

1   This, of course, becomes metaphysical in the cases of the discourses of cyberculture – discussion of being 'jacked-in' to the matrix, etc., plugged in to the medium.

2   In his treatment of Saussure, Derrida merely repeats the much earlier development that is offered by Jacques Lacan (1985).

3   This would imply a mode of ideality in which, for example, a sign could be infinitely repeated and withstand modification of its meaning, a project which occupied the philosopher Edmund Husserl.

4   For an extended review of Derrida's writings see Holmes (1989) and Norris (1982).

5   This is implicit throughout Derrida's work and is explicit in the opening of SEC with a distinction between non-semiolinguistic and semiolinguistic communication, between communication in the sense of bridging a gap, making close what was afar, and the sense of communication as the transmission of meaning.

6   Except that the medium actually provides such a context when engaged with ritually. It is precisely because the original context cannot be reproduced that we look to the medium itself.

7   But also its inadequacy compared to the civic virtues of print (see Marc, 2000: 637). Telephones have seldom met with the same kind of critique.

8   Alexander argues that news operates in a paradigm of truth-telling, but that when it addresses an issue concerned with crisis, this function gives way to being a forum for what he calls 'a certain ritualized value experience'. In a case study of the reporting of Watergate in the US media in 1973, Alexander argues that the media had become more concerned with '"values" that define the meaning of nation, the nature of citizenship, the duties of office, or the meaning of life itself. When truth is broadcast on such a generalized level, there is media ritual' (245).

9   See Regis (1990) for a discussion of how postmodern technoscience rejoices in exceeding and abjuring the body.

10   For an exemplary instance of this treatment of identity, see Donath (1999).

11   Using the example of television, Thompson illustrates this point (Thompson, 1995: 96).

12   Thus, as Thompson (1995) points out, it is possible to 'communicate through television' (92).

13   In a rare articulation of this form of integration that is conceptualized in the interests of the promotion of talk shows, Couldry cites a television producer from Dominique Mehl's analysis of intimacy and television, who explains the attractions of TV for self-disclosure. 'It's as if, in order to speak to those close to them, it's necessary [for them] to pass through TV. One could say that, in order that these people are reintroduced into the social circuit, they must pass through television. … Which is their home' (quotation translated in Couldry, 2003: 124).

14   Thus reciprocity does not require the 'logocentric' here and now of *interaction*, but is made possible by the *anticipation* that it is at least possible to be in the position of an interlocutor (as in the case of 'para-social interaction'). Or as Walter Ong (1982) says: 'In real human communication, the sender has to be not only in the sender position but also

in the receiver position before he or she can send anything. ... Human communication is never one-way. Always, it not only calls for response but is shaped in its very form and content by anticipated response' (176).

15 On the positive side are: investigative journalism, whistle-blowing, law enforcement, self-help, personal privacy protection, avoiding persecution (113–14). The negative includes: spamming, deception, hate mail, impersonation and misrepresentation, on-line financial fraud, numerous illegal activities (115–16).

16 These forms of communion may have positive and negative functions: they might enable a shared sense of belonging whilst also masking inequalities and conflicts within the social order.

17 'To the extent that "everything works *as if*" there were a functioning social whole, media and media rituals are central to that construction – which is why we need to study them' (10).

18 Thus Couldry wishes to reject theorists who argue in a limited or general sense that media rituals are extensions of other forms of everyday ritual. For an instance of the latter, and a comparative analysis of rites, ceremonies and media ritual, see Rothenbuhler (1998).

19 However, Couldry's book makes little attempt to theorize the relationship between mass media rituals and New Media rituals.

20 Couldry enlists the work of Italian political theorist Albert Melucci for this notion, but he could just as easily employ Slavoj Žižek's work in *The Sublime Object of Ideology* (1989).

21 For Couldry, the 'ritual' is not an activity, but marks an entire sphere of integration (2).

22 Significantly, Calhoun says, such tertiary relationships might involve ordinary written communication; they 'need not involve electronic technology, though such technology now greatly enhances the reach and the efficacy of such systems' (332).

23 Calhoun's tertiary and quaternary levels are dealt with in most CMC literature in terms of use/abuse, 'impact analysis' or within the sociology of technology in terms of a positive and negative effects debate (see, e.g., Spears and Lea, 1994).

24 A major work which configures such primary relations as a level of sociation is Bott (1971).

25 'Certainly, they think, a world dominated by relationships conducted over the phone, by correspondence, or with the assistance of computer would be much worse' (Calhoun, 1986: 335).

26 Sharp distinguishes between social form and societal form. Social forms are modes of integration that feature identifiable bases of community, virtual, extended, face-to-face, whereas a societal form refers to an actually existing historical 'configuration' of the different levels and components of social integration and their institutions (see Sharp, 1993: 225).

27 Slevin's caveat is as follows:

> It must be remembered, however, that the internet cannot be approached as a single communication entity. ... It consists of an array of different technical applications. A more detailed study would involve the examination of various internet applications, for example WWW, e-mail, IRC etc. and the unique way in which they combine Thompson's 'attributed of technical media'. (62–3)

SIX

TELECOMMUNITY

Rethinking community

For a term that is so over-used in media publics, it is remarkable how under-theorized 'community' is today. Since the nineteenth century, when Ferdinand Tönnies formulated what has become the most widely referenced understanding of community, in his *Community and Association* (1995), little formal analysis of community has been undertaken. And yet a certain regard for community has constantly endured throughout the discourses of modernity as a key term of reference and as a legitimating narrative for the human sciences and civic discourse.

For example, in the documentation of modern field research, 'community' is a key identifier for research into 'impact assessment'. The destination of such 'impacts', whether they are of electronic media, urban developments or just about any governmental policy it is possible to name, is invariably 'community'. And yet, oddly, few of these documents feel compelled to define 'community' at all. At best, they tend to defer to 'community' as a legitimating narrative which is safe to deploy precisely because of its ambiguity.

In the nineteenth century, the principal theorists of community were Tönnies and Émile Durkheim. Durkheim's concept of the *conscience collective* can be added to Tönnies' distinction between community and association, as foundational theories which have been little explored in terms of their relevance today. At the same time, new conceptions of community have co-emerged in relation to globalization and telecommunication which either reinforce mythological conceptions of community, by arguing that such a fiction is being 'lost', or advance new bases of human association that did not exist previously.

In this chapter, the relevance of the old and the new accounts of community to studying media and communications will be thoroughly examined. In particular, we will be looking at whether broadcast and networked mediums of communication can, in themselves, provide contexts for community as defined by these accounts. The different characteristics of these types of community will be outlined and their interrelationship will also be explored. But first, we will examine classical theories of community as well as some recent claims about the resurgence of community.

Classical theories of community

Until recently, conventional usages of the term 'community' in the human sciences had tended to render it as a formalization or deviation from what Émile Durkheim described as the *conscience collective*, which he defines as 'the set of beliefs and sentiments common to the average members of a single society [which] forms a determinate system that has its own life' (cited in Lukes, 1973: 4). This 'life of its own' is one which earns it the status of a 'social fact', having an effectivity and role to play in social integration. *Conscience collectives* are typically based on organized core values such as those exhibited by a religion. In such a case, a social group, or even entire societal forms, can be enveloped by a system of belief which becomes an overarching organizing mechanism of association.

Such an overwhelming centralized means of association, which Durkheim associated with traditional or 'mechanical' societies, was becoming attenuated by the advance of modern societies. Religion itself comes into crisis as the institution of the church becomes just one of a plurality of institutional sub-systems.

The move to what Durkheim called 'organic soldarity' is also marked by an increasing division of labour that becomes the organizing agent for social integration. The individualism inherent in specializing in a job becomes a basis for differentiation, which is itself elevated to a belief and a basis for a new kind of solidarity. Thus, for Durkheim, there is less stress on the *conscience collective* as being based on ideas, and more on the recognition of the importance of institutions, from family, to education, to workplace, and, at the same time, a recognition of the necessity of the division of labour.

A much neglected aspect of Durkheim's account of the *conscience collective* is his emphasis on the importance of material social facts: the institutions of society, population density, but also 'the number and nature of channels of communication'. The material, structural features of a society radically shape the forms of association which they can facilitate (Durkheim, 1982: 58).

As populations increase, and the urban architecture which they cohabit becomes more and more private, the so-called 'dynamic density' of society begins to change. In such conditions, the means of communication and transportation become vital to maintaining anything like the kind of community found in pre-industrial, pre-media kinds of societies. The forms associated with this kind of society – what Tönnies called *Gemeinschaft* – have, in media societies, all but hollowed out. Besides *Gemeinschaften* of religion, Tönnies lists *Gemeinschaften* of language, and of place, as the most important basis for such forms of solidarity (Tönnies, 1955). In the modern era of globalization, these forms are rapidly breaking down.

Gemeinschaft is a form of 'unity in plurality', it is close-knit: 'the intimate, private, exclusive living together – like a family'. This, Tönnies contrasts with *Gesellschaft*, which is a form of plurality in unity. *Gesellschaft* is public life – it is the world itself. 'One goes into *Gesellschaft* (society) as one goes into a strange country' (38).

In contrast to *Gemeinschaft*, *Gesellschaft* (society) is transitory and superficial. Whilst *Gemeinschaft*, should be understood as a living organism, *Gesellschaft* (society) [is] a mechanical aggregate and artifact' (39). *Gesellschaft* (association) is always by way of contractual arrangement, or cooperation towards a specific aim. *Gesellschaft* is well suited to Durkheim's principle of individualism. 'In *Gesellschaft* every person strives for that which is to his own advantage and affirms the actions of others only in so far as and as long as they can further his interest' (88).

For Durkheim, the onset of what Tönnies calls *Gesellschaft* created a weak overall sense of a *conscience collective* for any given society as a whole, and for him, the division of labour was not a sufficient unifying force to overcome the loss of an ideational bond. Instead, the *conscience collective* contracts to institutions, in which solidaristic attachments can become feverishly strong. It is as though, as compensation for the absence of any kind of overall integration into society, that 'strange country' which has now become the world itself, individuals seek refuge in the private and closed environments of institution and family. This 'miniaturization' of community can also realize itself in the workplace, subcultures and, as we shall see, television and the Internet (Fukuyama, 1999).

The change from traditional to modern societies is therefore a transformation in the architecture of community. In Durkheimian terms it does not mean that individuals in modern societies are more weakly integrated than they were in traditional societies; it is just that such integration is concentrated into sub-systems of the social order.

It is, however, true to say that in modern societies individuals do not look to the 'social whole' for a sense of integration, and generally feel a sense of anomie in relation to such an entity. But such a general condition of anomie is not a prescription for a romantic return to close-kit community, which communitarian movements espouse, a movement which was at its most salient in the nineteenth century, precisely when the bourgeois and industrial revolutions of Europe were experienced most bluntly and starkly.

Rather, unlike Tönnies, Durkheim saw the kind of solidarity exhibited by traditional societies as having its own particular problems, to do with over-integration and over-regulation. The close-knit community might well be intimate and highly connected, but it can also be suffocating, oppressive and imprisoning.

These problems of over-integration have today been transferred to those institutions which have miniaturized community – the workplace and the family being obvious ones. To take the workplace, the rise of workaholism, and the phenomenon of people living to work, rather than the other way around, can create enormous pressures, leading to permanent stress, depression, even suicide. Similarly, the modern nuclear family is experienced by many teenagers as too suffocating. They look to ways of escape, and when they do escape, they typically take all kinds of risks, as a means of rebellion and compensation for a life course of over-regulation. Paradoxically, the over-regulation that is found in institutionalized forms of community stands in an inverse relationship to the very weak kinds of tie that are found outside these protected bubbles of community. Whilst they may become suffocating, the cost of leaving them may be to accept a condition where contract and law must perpetually rule over mistrust and individual interest.

The contrast between the micro-community and the outside world grows ever more pronounced in modern/postmodern societies, to the point at which integration into the larger contexts of what Cooley called secondary relationships as well as tertiary, indirect mediated relationships becomes difficult for many people.

Another major aspect of this contrast is around the question of belief. As community miniaturizes, the number of settings in which individuals are enveloped by different belief systems also multiplies. Thus the discourses within the family will be different from those in the school yard and different again in the workplace. In most cases, these differences come into conflict with one another, and for an individual to cope with this fact s/he will invariably adopt different and contradictory subject positions. The fact that the theory of the subject emerged at the zenith of the period of modernity needs to be related to this question of the pluralization of settings of integration.

Such pluralization need not be seen as a dilution of community from the point of view of a given *individual*. Bell and Newby (1976) argue that community is only possible when the connection between individuals can be characterized as *multi-stranded*. These strands could be based in repetition of meeting in the street, kinship, membership of a group. Like Tönnies, Bell and Newby distinguish between three different forms of community: there are geographically based communities of propinquity, which do not necessarily require a strong *conscience collective*; there are communities of a localized social sub-system, such as in institutions; and, lastly, there is the populist sense of community as belonging and goodwill, which is described as *communion*. Communion need not depend only on parochial assembly and may well occur at a distance.

But for Bell and Newby, neither an extended nor a local sense of community is necessarily sufficient on its own. What is lacking in a technologically extended communication event might be supplemented by embodied travel. Conversely, the more that people travel in an embodied manner, the more they might feel the calling to 'stay connected' via electronic or other means to the places that have been visited. But, for the most part, in environments of virtual electronic community, we are permanently exposed to a propinquity with strangers: on the screens we immerse ourselves in; in gazing from our car at high speed; or in halls of large volumes of *flânerie* – the airport, the shopping mall, the tourist bubble. In the midst of such a maelstrom and proximity of strangers beating out familiar pathways of movement, to 'travel' weightlessly, whether corporeally or in the imagination, is to establish *networks* of communication in which communities have their life.

The 'end of the social' and the new discourse of community

The rise of the 'network society', together with the contentions that social ties and practices are of a much more capillary nature in modern society, and that anything like a 'social whole' can be effective in the integration of persons, has informed recent proclamations concerning the 'end of the social'.

In a 1996 article on 'The Death of the Social', Nikolas Rose argues that, with the assistance of postmodern theory, 'the object "society", in the sense that began to be accorded to it in the nineteenth century (the sum of the bonds of relations between individuals and events – economic, moral, political – within a more or less bounded territory governed by its own laws) has begun to lose its self-evidence' (Rose, 1996: 328). For Rose, '"The social" … within a limited geographical and temporal field, set the terms for the way in which human intellectual, political and moral authorities, in certain places and contexts, thought about and acted on their collective experience' (329) – which is, for Rose – the nation-state.[1]

Similarly, Alain Touraine (1998), in a account of the 'end of *Homo Sociologicus*', proclaims:

> We have learned to do without the idea of society as it was defined by ratio-
> nalist thought from the 16th to the 18th century, and as it was renovated
> and reinforced by the theorists of modernity, of industrial society, of the wel-
> fare state and also of national development policies. We have come to
> the end of the road to which the founding fathers of sociology led the way
> a century ago. (127)

For Touraine, society is neither a 'state of nature' nor a progressive framework of human development; rather, it has become a technology of managing populations which has recently exhausted itself. Since the 1970s, in texts like *The Self-Producing Society*, *The Voice and the Eye* and *The Return of the Actor*, Touraine has promoted the idea of a 'programmed society', namely that advanced industrial societies have developed 'the capacity to choose their organization, their values, and their processes of change without having to legitimate these choices by making them conform to natural or historical laws' (Touraine, 1988: 40).

The problem with such a society is that it produces a setting in which norms are rapidly changing because they are constantly being redefined, generating a crisis for how individuals are integrated. Individuals whose roles were once highly defined, what Touraine calls 'actors', must increasingly become more self-forming and self-active, without the programmed and governmental contexts which might give either 'instrumental' or 'value-rational' kinds of action any kind of solid meaning. Without roles to protect them, individuals are much more vulnerable when they go out into society (*Gesellschaft*), as the links between actor and system become diluted.

Like Touraine, Rose (1996) contends that '"[t]he Social" ... no longer represents an external existential sphere of human sociality' (329). Rather, within the modern nation-state, 'the social', as a project, is said to be replaced by what he calls a 'government of individuation', a form of subjection and control in which individuals are encouraged to be responsible for themselves and police their own behaviour and that of others.

For Rose, the development of social ontologies in the present age has been enveloped by discourses, ways of speaking and thinking which manage the experience of social reality. This is sometimes called, following Michel Foucault, 'governmentality'. The social only has a unity insofar as it is '*in the name of the social*' that various political pressures are said to be exerted on populations for the purposes of national governance. Many of these tendencies have become necessary, according to Rose, precisely because of the erosion of traditional state power by globalization, requiring a 'new spatialization of government' to manage politics and economics. This has brought about new kinds of discourses of control, in which abstract economic processes are mischievously spoken about in terms of communities of interest.

Such a language reterritorializes populations as groups of specialized markets within economic relations that 'do not respect national political boundaries' (Rose, 1996: 330). Rose asks: by what terminology are economic relations now understood, and how is economic governance posed, in the era of globalization? 'Consider the prominence of the language of community' (331).

Rose (1996) contends that the language of community has become a burgeoning terminology of political life and that it has replaced 'the social' as the centre of governmentality (see also Touraine, 1998). The most notable of these is the globalization of community, in which nationally constituted 'imagined communities' co-exist in narrative form. At the same time, nationalism itself is attenuated as the number of narrative identifications with community proliferates *ad nauseum*: terms as wide-ranging as, for example, 'the business community', 'the sporting community', 'gambling communities' – in fact the kind and range of divisions are almost limitless. The very discursive prominence of community is posited as proof enough of the reterritorialization of older (mostly geographic and ethnic) frames of belonging based on what Rose (1996) calls 'other spatializations: blood and territory; race and religion; town, region and nation' (329).

In the same way as populations are coming to be *discursively* divided into smaller and smaller distinct communities, they are also, in the opposite direction, 'called up' to communion with quite abstract kinds of global communities – virtual communities, the Olympic community, the 'international community'.[2]

Globalization and social context

For Rose, 'society' and community are viewed as discursive constructs that have changed roles. 'The formation of the notion of a national economy was a key condition for the separation out of a distinct social domain' (Rose, 1996: 337). Whereas, for Rose (1996), 'the social' once acted as a discursive agent for the integration of persons on the basis of 'social protection, social justice, social rights and social solidarity' (329), for Touraine (1998), the decline of national communities derives from the 'decomposition … of society' by way of the 'growing autonomy of the economic sphere from institutional controls' which 'exist in general at the national level' (129). Touraine acknowledges, contra the extreme globalists, that the events that are said to have produced globalization – 'the increase in international trade; the more rapid intensification of financial flows; the rise of new industrial countries; the birth of the information society … the end of the Cold War and the fall of the Soviet empire' – are not systemically linked, but rather are phenomena 'largely independent of one another, and are of different natures' (129–30).

Nevertheless, taken together, these discrete events are said to have social and political consequences, the first being the 'breakdown of social and political constraints on economic activity', producing 'the most radical rupture ever observed between actor and system', whilst the second is the weakening of the nation-state (130–1). For Touraine, the most significant change is in the decomposition of the institutionalization of norms in the social world: 'the main fact is that we no longer recognize the presence of norms in many realms of life' (131). Instead, as larger and larger spheres of behaviour are no longer said to be subject to nation-state/society-framed norms, '[t]he system is no longer a social one, but becomes a global market, self-regulated by law firms, rating agencies and international financial institutions and the financial markets themselves … the social actor disappears, and the remaining actors are no longer social' (130).

In the contemporary context, therefore, Touraine poses the main task for sociology today as having to discover 'a new principle, capable of replacing the idea of society and more specifically of national society, which for so long played this role of mediation and integration' (133). This leaves him with the question: 'how can we be actors and create space for autonomy between the globalized economy and communal cultures, neither one of which leaves room for the actor?' (135).

Touraine's own answer to this question is that 'there are no longer transcendent universal values that might unite all of humanity' (136), as in the case of the grand narratives of modernity; rather, capitalism, in its contemporary form, promotes individuation and difference, and it is this right to individuation and difference which is left over as the last viable universal value that is able to provide a normativity in contemporary life. Promoting Foucault's return to the question of the subject in his later work, Touraine turns attention away from a sociology of system and actor to that of the subject: 'The recognition of each person's right and capacity to become a subject is a universalistic value; as is the right to combine a commonly shared scientific or technological rationality and a particular cultural identity' (136).

Touraine's turn to a social theory of the subject through individuation, which he claims is necessitated by the global market's decomposition of social norms, shares much ground with the work of Jean-Luc Nancy in *The Inoperative Community* (1991) and Georgio Agamben in *The Coming Community* (1993). In these texts a philosophical rethinking of community is attempted in ways which are more suited to the fragmentations of modernity that are evident today. According to Agamben and Nancy, we can no longer speak of a transcendent principle or context of community other than the fact that subjects must be self-active in attributing any kind of global significance to their experience. To impose and presuppose community in the name of the transcendent is to disregard it since, as Nancy (1991: xxxviii) puts it, community cannot be presupposed and the thinking of community as essence is the closing off of the political. Community is realized in the very retreat from an organizing principle, the refusal of a universalizing essence.

Like Nancy and Agamben, Touraine (1998) does not appeal to a universalist discourse: 'while dominated groups used to refer to a meta-social principle – God, reason, history or the nation – in order to challenge the dominant group's power, today the defence of the subject invokes no higher principle and does not seek to obtain power' (138). Instead, for Touraine, the subject only struggles against the (now global) economic forces which are constantly threatening a reduction of being 'to a series of life experiences, resembling the television programmes one sees when one zaps from channel to channel' (136), while 'community' no longer has internal conditions but is formed and acts through strategies akin to Rose's account of governmentality.

The rise of global communities of practice

The governmentality perspective, which proposes that discursive strategies become increasingly important for the maintenance of nation-state forms of society, identifies globalization as the basis of the breakdown between system and actor. Globalization is seen to erode the middle-level agencies of social integration which were once provided by technocratic society.[3]

In the wake of such erosion, community, on the one hand, retreats to micro-communities, and, on the other hand, reaches out to more and more global forms of cosmopolitan integration. Such a re-spatialization of community makes it very difficult for Tönnies' *Gemeinschaft* or even the imagined community of the nation-state (Anderson, 1983) to persist.[4]

In a review of the different definitions of community that have appeared over the last fifty years, Ken Dempsey (1998) points to two characteristics which keep returning: first, having the social ties in common that produce a high degree of social solidarity (a structural characteristic); and, second, the experience of belonging together. As Dempsey points out, problems begin when we insist that both characteristics have to be present in order for community to exist:

> ... it is possible for the objective (or structural) characteristics of community to be present and the subjective characteristics to be absent. People may be linked by social ties of interdependence and yet have no sense of belonging together. It is also possible for the opposite to be true: for people who do not know one another, to have a sense of belonging together. (141)

The fact that the objective and subjective components of community are rarely in alignment, in a globalized world in which the relationship between language, religion and place is becoming increasingly arbitrary and cosmopolitan, suggests that other bases for community can come to the fore. From a Durkheimian point of view, all individuals need to be socially integrated in some way or another. How this occurs may vary enormously between individuals and according to the place in which they live, including the new 'places' that are able to come into existence, such as cyberspace.

However, it is increasingly clear in media societies that tradition and belief, or a *conscience collective*, are no longer an organizing basis for community. Through mediums and rituals, it has become quite orthodox for people who do not know one another to have a sense of belonging together in a mediated ceremony. In addition, the advent of cyberspace introduces the prospect that communities of place are not just geographical, and it also facilitates the possibility of meeting nearly everyone who has the same interest as you and is also connected to the Internet, wherever they are located.

However, the content of beliefs and interests is only one component of Durkheim's original description of the *conscience collective*. He also specified intensity – the intimacy of interactions, volume, the number of people enveloped by the interactions – and rigidity – the regularity and adherence of these interactions. This also means that social integration is very much based on the *practice* of interaction, not just on what it signifies. When we routinize our interaction with others and with the mediums through which we conduct such interactions, we create a world around us which becomes very familiar to us, regardless of what the content of our interactions is. Rituals are concerned more with this routinization than they are with trying to identify with the same ideas or aesthetic effect.

We can draw from Durkheim the observation that community is as much about practice as it is about belief, a distinction also recently made by Nancy Baym (2000) in her innovative analysis of the difference between 'audience community and network community'.

> Audience communities and online communities co-opt mass media for inter-personal uses. Grappling with the social nature of these new types of community requires understanding them not just as online communities (organized through a network) or as audience communities (organized around a text) but also as communities of practice organized, like all communities, through habitualized ways of acting. (4)

Whilst audience communities might be organized around image, music and text, both on-line communities *and* audience involve regularized forms of practice.

Drawing on formulations from Hanks (1996) and Lave and Wenger (1991), the communities-as-practice approach provides an instructive means of steering our way through the complex differences between broadcast and network forms of community, and without the sentimentalism so often ascribed to the term.

> At the center of the practice approach is the assumption that a community's structures are instantiated and recreated in habitual and recurrent ways of acting or practices. When people engage in the ordinary activities that constitute their daily lives, they are participating 'in an activity system about which participants share understandings concerning what they're doing and what that means in their lives and for their communities' (Lave and Wenger, 1991, p. 98). In short, if one wants to understand a community, then one should look to the ordinary activities of its participants. This is a fairly minimalist definition of community, without the warm and fuzzy connotations that many link to the term, but it is a definition that provides a workable core. Without shared engagement in a project, there can be no warmth and fuzziness. (Baym, 2000: 22)

Of the different indices of community, place, religion, language and ethnicity can be associated with communities of belief, whereas it is space and place which stand out as a site of practice.

As communities of belief become disassociated from particular places, owing to global movements of culture, modern communication becomes all the more important in order to sustain them, as they regularly occur over ever-greater distances.

However, at the same time, such means of communication make possible new kinds of spaces that are available to be practised. In this connection, the implications of Michel de Certeau's formulation that 'space is a practiced place' (de Certeau, 1988: 117) are far-reaching in their consequences. In practising places, it makes no difference whether they are global or local senses of place, corporeal or electronic spaces. Indeed 'cyberspace' is a place that we can become just as attached to by way of our routines of navigation as we can by the repetition of meeting people when we go into the street.

As I have argued in *Virtual Globalization* (Holmes, 2001), de Certeau's formulation is a very useful one for thinking across physical and virtual spaces. In *The Practice of Everyday Life* (1988), de Certeau explores how we can relate to our physical environment by practising routines of traversal in our daily existence and observation which can equally be applied to how individuals navigate media spaces.[5] Any medium which can extend experience necessarily brings together local and global kinds of settings, to the extent that it is globally accessible or extant. Indeed, as local worlds become subject to accelerating flows of messages, bodies, styles and commodities which course through them, attachment to electronic global spaces can become more attractive, to the extent that they offer a stable uniformity that can no longer be found on a local basis.[6]

Sociality with mediums/sociality with objects

The centrality of 'communities of practice' for understanding the inter-relation between local and global community also requires a differentia-tion between the corporeal and virtual means of engaging with such spaces. When we turn to the question of media spaces, there are two dimensions of interaction involved in all media: interaction with medi-ums; and sociality with the media technologies which give us gateways to such mediums.

As we saw in the previous chapter, interacting with mediums is explored by ritual perspectives of communication, whereas intersubjec-tive communication is explored by transmission accounts.

The argument that, in information societies, individuals increasingly interact with a medium rather than with other interlocutors is one which is well supported not only by ritual communication views but also by neo-McLuhanist and abstraction views. Moreover, the fact of interacting with mediums is apparent whether we are discussing broadcast or network integration.

Whether we choose to single out television or the Internet, interact-ing with mediums tends to have two consequences: ritual forms of attach-ment to an electronic assembly are usually accompanied by a parallel tendency to fetishize the personalization of media technologies. In the case of mediums, it is not that they 'mediate' our relationship to others; rather, it is the medium itself which is worshipped. In the case of the per-sonal media technologies, the relationship to objects with which commu-nication is enabled can become more intense than to other persons.

Sociality with media

As identified in the previous chapter, medium theory has a habit of differentiating social forms according to regimes of communication. Further, it tends to inflate the technical importance of mediums to the point where a given era is defined by the technology itself. In this it does not consider the question of social dependence on mediums, and the question of social determination, which structures the way in which some mediums become dominant over others. Moreover, medium theory tends to argue that one rather than another medium is dominant because more people are physically interacting with it. Integration theory, on the other hand, argues that in any given communicative situation there is always a co-presence of levels of mediatized integration involved – face-to-face, agency-extended, etc. – but that one of these levels can come to recast how all other levels are experienced. Thus, for example, in modern society it is possible for those immersed in technologically extended communication cultures to engage in face-to-face relationships in a disembodied register. This is not to point to the fact that the modern 'computer nerd' engages little in the way of eye contact. It is rather that such subjects necessarily adopt CMC as a frame of reference, even when they are not engaged in such communication. Conversely, persons formed in cultural settings where the face-to-face relationship is dominant will tend to attribute everything with a face-to-face character, even when they are not actually engaged in mutual presence.

The integration approach to media is distinguishable from interaction approaches which confine themselves to intersubjectivity. As we saw in the previous chapter, integration occurs not by interacting with others, but by interacting at the level of an indirect social relationship (Calhoun, 1992), one that is extended by another agent, be it another person or a communication technology.

Moreover, following Meyrowitz, communication mediums possess the quality that they can extend experience in ways that exceed the kinds of behaviour found in local environments: media are not merely 'channels for conveying information between two or more environments, but rather shapers of new environments themselves' (Meyrowitz, 1994: 51). In a more recent account, 'The Shifting Worlds of Strangers', Meyrowitz (1997) argues that such re-territorialization changes the relationship between 'them' and 'us'. The balance between oral, print and electronic environments profoundly changes the relationship between strangers and 'familiars'.

In regarding media as environments of sociality, traditional charges of 'technological determinism' are no longer pertinent in the way they were in the earlier years of media studies. One of the most cogent and persistent accounts of technological determinism is that of Raymond Williams in his *Towards 2000* (1983) and *Television: Technology and Cultural Form* (1974). In these texts Williams is very critical of the way in which one particular technology could be said to carry a spirit or *Geist which defines an age*. Secondly, he rightly points out that few analyses relate the *causes* of the development of individual technologies to a range of locatable political and economic interests, rather than simply technical progress. For him, technological determinism is an outcome of some or other 'expressive totality' view of technology as inevitable, and he wants to restore a sense of agency and policy to media technology.

However, Williams does not address the fact that when media become apparatus and networks which people live with and work with in an everyday sense, they have meta-psychological and social dynamics which are quite independent from how they have emerged. Of course the tendency to place the meaning of these networks and apparatuses within some kind of expressive *Geist* theory is a simplifying temptation – a simplification which Williams is justifiably annoyed with.

The kind of sociality which emerges in media environments is typically 'embedded' in networks or groupings of techno-social mobility. These networks, which *always* involve actions that are embedded within technical means of communication or transportation, become meaningful-in-themselves, *not* as means to extend face-to-face relationships.

John Urry (2002) argues that in most contemporary literature on mobility an appreciation of the embeddedness of forms of sociality in a network/medium is precluded by the pervasive dichotomies within which technologically extended mobility is thought. The dichotomies of '[r]eal/unreal, face-to-face/life on the screen, immobile/mobile, community/virtual and presence/absence' (7) are each premised on the former term providing some pre-virtual knowable other which is in some way threatened or transformed by new forms of mobile 'at-a-distance' connections.

> Normally, this knowable other is characterised as 'real', as opposed to the airy, fragile, and virtual relationships of the electronic. And the real is normally taken to involve the concept of 'community'. Real life is seen to comprise enduring, face-to-face, communitarian connection, while the virtual world is made up of fragile, mobile, airy and inchoate connections. (2002: 1)

In opposition to such a limit-paradigm, Urry argues that all societies have exhibited and continue to exhibit diverse 'at-a-distance' connections, 'more or less intense, more or less mobile, and more or less machinic. ... All social relationships involve complex patterns of immediate presence and intermittent absence at-a-distance' (1).

By arguing that extended forms of connection are just as 'real' as propinquitous ones, or that, in a sense, all forms of community are 'telecommunities', and vice versa, we avoid one-dimensional, utopian or dystopian accounts of community.

That persons may become attached and embedded in techno-social networks across vast dimensions of metropolises and the globe does not mean that such networks have abolished less technologically mediated networks; indeed, they are typically mutually constitutive. Where technical embeddedness becomes more and more normative, other forms of connection like the face-to-face often become revalued. And with this form of connection, direct interaction with another person can become substituted by interaction with media objects.

Sociality with objects

Dependence on techno-social networks may exhibit different kinds of intensity depending on what other kinds of connections and circulations co-exist with them. The more embedded are technologies in such networks, the more it is possible to engage with these technologies as *ends in themselves*. To the extent that such embeddedness creates a techno-social way of life, 'transport' models of understanding communicative events become redundant, as is the social type said to be at the centre of this model, the media technology 'user'. User perspectives (see, e.g., Silverstone, 1991; Silverstone et al., 1991, 1992) are typically interested in which technologies are taken up by particular individuals and the purposes for which they are used, but do not examine the interaction between individuals and objects, nor how such interaction can actually alter the identity of the person interacting. What the user perspective also ignores is that *objects of interaction* are means of maintaining the connection to mediums/networks.[7]

The television, the Walkman, the mobile phone, the motor car, the keyboard – people become *very* attached to these commodities, and begin to relate to objects rather than to other people. The greatest fetishism of commodities is of these means of connection. Relations to other persons become refracted through these objects, or they may become confused with our own narcissistic relationship to the technology itself.

In a longitudinal study (1991–6) of over 400 TV audience diarists in the UK, Gauntlett and Hill (1999) document the companionship which the TV set provided for viewers of all age demographics. 'When respondents wrote about what television meant to them, they often listed the information and entertainment aspects of television, but mentioned as well the company and even "friendship" that it offers' (115). Two kinds of attachment were reported: the TV set itself as 'friend' or a member of the family; and TV as bringing 'friends' from outside the home (115–19). Many of the diarists who reported strong emotional affections toward their set lived alone. But equally, many reported some degree of guilt about issues such as daytime TV viewing being a waste of time, or being selfish in watching a programme which the viewer knew that few family members, friends or visitors would be interested in.

Arguably, the more *personalized* a communication technology, the greater the human–technical interface. McLuhan, in a chapter in *Understanding Media* (1994) entitled 'The Gadget Lover', argues that in media societies, individuals everywhere encounter themselves in a world which becomes closed and virtualized, which, as the story of Narcissus tells us, requires a fascination with media which extend this closed bubble in which others become resolved into our own image.

> The youth Narcissus [from the Greek word *narcosis* or numbing] mistook his own reflection in the water for another person. This extension of himself by mirror numbed his perceptions until he became the servomechanism of his own extended or repeated image. The nymph Echo tried to win his love with fragments of his own speech, but in vain. He was numb. He had adapted to his extension of himself and had become a closed system.
>
> Now the point of this myth is the fact that men at once become fascinated by any extension of themselves in any material other than themselves. (1994: 51)

Medium theory argues that once technologies are integrated into a 'way of life', it may be difficult to be without them. Indeed, an individual's intolerance at having a connection broken gives a precise measure of how attached he or she is to that medium. It also changes our very conception of what a medium is. McLuhan, for example, claimed that electric light and the motor car are both mediums. It is certainly true that if either stopped working in our immediate environment most of us would seek to rectify this problem very quickly to the extent that we are very dependent on them. In the case of transport, the centrality of travel in sustaining community networks is never more visible than when it breaks down, be this a public or private instance. Typically, a train strike or a flat battery is thought of purely in terms of inconvenience. However, the propensity for individuals to treat such events as *moral* crises suggests that they are significant in ways much more to do with community.

In circumstances where our everyday actions become embedded in technological networks, technology itself becomes transparent. The philosopher Martin Heidegger pointed this out in his analysis of the way in which technology destines the world to be revealed as a reserve of utility (see Heidegger, 1997). When technology as 'equipment' is routinely used to achieve given ends, the power that it holds can become taken for granted as the technology itself becomes invisible. As Knorr-Cetina (1997) explains of Heidegger: 'Equipment becomes problematic only when it is unavailable, when it malfunctions or when it temporarily breaks down. Only then do we go from "absorbed coping" to "envisaging", "deliberate coping" and to the scientific stance of theoretical reflection of the properties of entities' (10).

The implications of Heidegger's insight into the conditions of the visiblity/invisibility of technology in everyday life are integrated by Knorr-Cetina into an extremely novel account of the social bonds which individuals form with technological objects.

Given that individuals must develop a certain 'technical intelligence' in order to cope with technological change, with switching between the use of different technologies, or to overcome the malfunction or dysfunction of technology, Heidegger argues that technological society presupposes a more 'theoretical attitude' oriented towards objects of knowledge, rather than 'practical reason', which is concerned only with instruments. Thus, what was once the specialization of science and experts becomes an 'attitude' throughout the population.

Such a change prompts Knorr-Cetina to examine the relation that scientists and experts have to objects in developing an extended account of what she calls 'objectualization'. Objectualization describes the way that 'objects displace human beings as relationship partners and embedding environments, or that they increasingly mediate human relationships, making the latter dependent on the other former' (1).

For Knorr-Cetina, there are two principal driving forces of objectualization: 'The first is the spread of expert contexts and knowledge cultures throughout society that discharge these cultures into society as a possible driving force behind the rise of an object-centred sociality' (23). The second are the 'relationship risks' that many find inherent in contemporary human relationships.

Where human relationships fail, individuals turn to objectual relationships as compensation. Knorr-Cetina describes this as a post-social development, where 'social' is reserved for forms of societies based upon *solidarity* in the Durkheimian sense (18), the unity of something shared, the unity of a moral field or a unity of meaning.

Like Touraine and Rose, Knorr-Cetina argues that information societies are undergoing a post-social transition. But post-social transitions imply that social forms as we knew them have become flattened, narrowed and thinned out; they imply that the social is retracting. Knorr-Cetina points out that this is usually explained in Durkheimian terms as a further boost to individualization and the loss of a common culture (6).

> ... as common values are no longer at the award growth of shared traditions and cannot just be imposed by some authority, integration via norms and values appears to be less and less effective. In fact, this sort of integration is imaginable today only as a socio-culturally engineered consensus. (24)

But the shift away from such value consensus and the rise of individualization need not be read as the 'death of the social': '*postsocial relations are not a-social or non-social*' (7). Rather, Knorr-Cetina argues that it is an error to characterize individualization in terms of 'human relationships' in the present period. If we take into account the ways in which human beings tie themselves to object-worlds, then the ordinary concept of individualization is problematized. In that case, 'objects may simply be the race winners of human relationship risks and failures, and of the larger postsocial developments' (23) In other words, for Knorr-Cetina, 'Individualization intertwines with objectualization – with an increasing of our orientation toward objects as sources of the self, of relational intimacy, of shared subjectivity and of social integration' (9).

However, for Knorr-Cetina, the relationship of individuals to technology isn't just a matter of ritual, although this is strong, but of turning everyday objects like commodities and instruments into knowledge objects in ways that become meaningful as such. Where the propensity of a particular individual to do this (which may vary according to gender and age) is either resisted or too intense, it attracts a label – technophobia or addiction (see Brosnan, 1998).

In Knorr-Cetina's work, where the notion of solidarity is brought into the scenario of an object-centred sociality, it also needs to be epistemically grounded and not only ritually derived (19). In other words, having an epistemic intimacy with the object is important, not just the routine ritual of 'using' it as a means to some end. And indeed, in post-social societies, Knorr-Cetina claims that such an intimacy is beginning to eclipse intimacy with other human beings.

Consider the following statement from William Mitchell (1996) about his personal use of the Internet:

> The keyboard is my café. Each morning I turn to some nearby machine – my modest personal computer at home, a more powerful workstation in one of the offices or laboratories that I frequent, or a laptop in a hotel room – to log into electronic mail. I click on an icon to open an 'inbox' filled with messages from round the world – replies to technical questions, queries for me to answer, drafts of papers, submissions of student work, appointments, travel and meeting arrangements, bits of business, greetings, reminders, chitchat, gossip, complaints, tips, jokes, flirtation. I type replies immediately, then drop them into an 'outbox,' from which they are forwarded automatically to the appropriate destinations. If I have time before I finish gulping my coffee, I also check the wire services and a couple of specialized news services to which I subscribe, then glance at the latest weather report. This ritual is repeated whenever I have a spare moment during the day. (7)

What emerges here is that, whilst the meetings, the travel, the 'content', the business of institutional life vary from day to day, the technological ritual of clicking on is constant, enduring and cathartic.

Digital intimacy

Theories of object-relations are not entirely new, and have had a long-standing tradition in psychoanalytic theory. But few have related such theories to a transition from one kind of social form to another, or to the significance of media as relationship partners. One such theorist, referred to by Knorr-Cetina, is Sherry Turkle, who manages to combine both a concern for social change and an attachment to digital technology.

In *The Second Self* (1984) and *Life on the Screen* (1995) Turkle is centrally concerned with the self–other relations in New Media environments. In her earlier work, Turkle (1984) explores the way in which many computer users relate to computers as though they have a mind and conversely view themselves as machines. For Turkle it is the very opacity of digital technologies, the fact that, if you were to open one up you would not see tangible moving parts, gears, levers or wheels, but rather, simply, 'wires and one black chip' (22), that encourages us to speak of them in psychological terms. The computer is an evocative object with which we can develop an almost spiritual relationship (306): 'For adults as well as children, computers, reactive and interactive, offer companionship. They seduce because they provide a chance to be in complete control, but they can trap people into an infatuation with control, with building one's own private world' (19).

For Turkle, making the computer into a second self, finding a soul in the machine, can substitute for human relationships, a path dependence which can come to be both cause and effect of a new kind of hysteria:

> Terrified of being alone, yet afraid of intimacy, we experience widespread feelings of emptiness, of disconnection, of the unreality of self. And here the computer, a companion without emotional demands, offers a compromise. You can be a loner, but never alone. You can interact, but need never feel vulnerable to another person. (307)

In her later work, Turkle (1995) turns away from the object-narcissism of the PC to the decentred and multiple identities of the Internet. In cyberspatial worlds, rather than the virtual world of face-to-screen, it is possible to reveal ourselves in new ways. The Internet has achieved in practice what psychoanalysts have long been trying to achieve in theory: the realization that the autonomous ego is a fiction (see Turkle, 1995: 15). With much relief, Turkle revisits the insights of psychoanalysts whom she studied many decades earlier, simply by 'tinkering' with the Net:

> In my computer-mediated worlds, the self is multiple, fluid, and constituted in interaction with machine connections, it is made and transformed by language; sexual congress is an exchange of signifiers; and an understanding follows from navigation and tinkering rather than analysis. And in the machine-generated world of MUDs I meet characters who put me in a new relationship with my own identity. (15)

Media equations

From an entirely different direction, and without recourse to psychoanalytic theory, it is significant that media–object relationships are also being researched by computer corporations. Microsoft commissioned one such study, 'Social Responses to Communication Technologies', written by Reeves and Nass (1996). In their *The Media Equation: How People Treat Computers, Television, and New Media Like Real People and Places,* they declare, triumphantly but simplistically, that *media equal real life* (!): 'We have found that individuals' interactions with computers, television, and new media are *fundamentally social and natural,* just like interactions in real life' (5).

The simplistic import of their 'equation' is that 'the social' is defined in the narrow, everyday sense of 'politeness'. Nevertheless, their observations concerning interaction bear some attention despite the fact that they lack contextualization in any recognizable social analysis. The social responses study indicates that people are polite to well-designed computers, that screen motion draws similar responses to real motion, and that, psychologically, object-relations to TV screens are not so different from those to PC screens.

Mostly, these relationships are of a passive order; they 'do not apply to the rare occasions when people yell at a television or plead with a computer' (253). Interactions between persons and media technology lead individuals to 'allocate attention, assess competence' and organize information in ways which aren't just about efficiency or entertainment (253).

Such research, which is also the subject of numerous studies on mobile phone use in Katz and Aakhus's collection *Perpetual Contact* (2001), can sometimes lead to New Age kinds of spiritualism represented in attempts to suggest a new kind of community technospirit which emerges within a particular medium.

Katz and Aakhus, in their sixteen-page review of the essays in *Perpetual Contact,* feel compelled to invent a new term for such a spirit which emerges out of mobile phone use, which they call *'Apparatgeist'* (2001). In advancing this term, the authors rapidly slide from a concern with (as the essay title suggests) 'The Meaning of Mobile Phones' to universal claims about history, society and the human spirit! The tendency for New Media theorists to coin neologisms to define a new epoch, as we saw with the second media age thesis, or to make such grand universal claims, a matter to which I will return, is interesting in itself. For Katz and Aakhus, the term *Apparatgeist* achieves no less than to 'tie together the individual and collective aspects of societal behaviour' (307). Having defined *Apparat* as a term that can be found in numerous dictionaries and then defined *Geist* as a handy word derived from Hegelian philosophy, they see their new term as suitable for referring to 'the common set of strategies or principles of reasoning about technology evident in the identifiable, consistent and generalized patterns of technological advancement throughout history' (307).

Apparatgeist is the 'master concept which is informed by a 'logic' which is called 'perpetual contact'. In turn 'perpetual contact is a socio-logic' of 'personal communication technology' or PCT (307). From there, Katz and Aakhus run through a series of random but no less grand theoretical consequences of how 'PCTs' make up the *Apparatgeist*:

- They have a *Geist* which can be likened to the expansion of freedom.
- They have their own logic that informs the judgements people make about the utility or value of the technologies in their environment.
- They inform the predictions that scientists and technology producers might make about personal technologies.
- They also have a socio-logic that results from 'communities of people "thinking and acting together over time"' (307).
- In making possible *Apparatgeist* PCTs, 'the compelling image of perpetual contact is the image of pure communication … which is an idealization of communication committed to the prospect of sharing one's mind with another, like the talk of angels that occurs without the contraints of the body' (307).

Moreover, perpetual contact has a univeral historical status, parallel with the 'the image of perpetual motion that has driven the machinery of the past two millennia' (307). 'Whereas the idea of perpetual motion concerns the means of production, perpetual contact concerns the means to communicate and interact socially, which is fundamental to humans' (308).

Whilst Katz and Aakhus want to stress that their term does not require technological determinism, because it is about constraint of possibilities (307), their account is technological determinism at its worst, as well as tautological. The essay is subtitled 'A Theory of *Apparatgeist*', a theory of something that is predicated on their own theory. Their inability to even imagine that any of the phenomena in the list above, even the more credible amongst them, are attributable to any agents other than PCTs and their *Geist* is worrying, but unfortunately typical of a growing methodological essentialism around New Media impacts.

Post-social society and the generational divide

> Youth instinctively understands the present environment – the electric drama. It lives it mythically and in depth. This is the reason for the great alienation between generations. Wars, revolutions, civil uprisings are interfaces within the new environments created by electric informational media. (McLuhan and Fiore, 1967: 9–10)

A key consequence of the transition to an information or knowledge society that has received a lot of attention is the widening of the differences between generations accompanied by a contraction of the time in which this gap appears. Moreover, New Media environments such as 'cyberspace' are said to have their own 'time-worlds' which operate at far greater cycles than other forms of time.[8] In shorter and shorter cycles, the way in which persons are formed by media is quickly outdated. Computer companies employ adolescent 'geniuses' who seemingly have a natural aptitude for computers, and form their own networks of association which older generations cannot understand.

Some writers, such as Mark Dery, suggest that this is the basis of an epochal change, the decisive process which justifies something approximating a second media age thesis. Increasingly, computer culture, or cyberculture, seems as if it is on the verge of attaining escape velocity in the philosophical as well as the technological sense' (Dery, 1996: 3). For those immersed in it, cyberculture 'resounds with transcendentalist fantasies of breaking free from limits of any sort, metaphysical as well as physical' (p 8). This, it is argued, leads to a breakdown of cultural transmission and a separation of world-views.

It is certainly true that youth lead the way in their take-up and consumption of New Media, with older age groups perpetually catching up. In Finland, for example, which has the highest density of mobile phone use in the world – 64% at the beginning of 2000 – figures clearly suggest that take-up of the technology has been highest among the two youngest groups, and it steadily declines in proportion to age group. As at 2000, personal ownership statistics are revealing for the following age groups 15–19 (77%), 20–9 (86%), 30–9 (77%), 40–9 (67%), 50–9 (59%) and over 60 (29%) (see Puro, 2001: 21).

In Finland and in Norway, which has the second highest mobile phone density, teenagers stand out spectacularly as the highest usergroup of SMS (short messaging services, or 'texting'). For teens, it is associated with extensive subcultures. As Skog (2001) has remarked of the Norwegian experience, 'SMS has spurred teens to create an anglicized clique-based abbreviated language' (262). According to Skog's statistics, 75% of girls and 62% of boys regard SMS as an essential feature of their phone (262).

The fact that a technologically mobile network of youth are able to develop a highly specialized and defined communication culture that excludes nearly all other age groups illustrates the potential rigidities of the digital divide.

The ease with which armies of 'cyberbrats' are able to take up New Media stands in stark contrast to the 'technophobia' with which older generations are often tinged. Technophobia is not an opposition to technology-in-general, but a fear of the technologically very new and of the pace in which this newness colonizes the life world.

The youth-biased take-up of New Media does not, in itself, demonstrate resistance among older age groups. Rather, it is more likely that younger people will find New Media intuitive because they do not have to adapt from a prior regime of working with media apparatuses. When young people learn computers, for example, their complexity rapidly becomes transparent. In Heideggerian terms, children already have a theoretical attitude in working with computers, for which practical reason is naturalized. However, for those without this attitude, significant anxieties may be aroused in using any kind of technology.

Brosnan (1998), in his positivist psychology of *technophobia*, suggests that 'until technology becomes invisible, it will be found to create feelings of anxiety within certain individuals' (2). In the interests of novelty, Brosnan tries to suggest that young people may suffer from technophobia on account of greater expectation over their proficiency. However, this argument is not supported by the studies. What is important, though, is that older people who live alone tend not to be so 'technophobic', possibly because they may welcome object-centred relationships as substitutes for human relationship partners who have been lost. Studies in Germany of older age group television audiences display a similar trend (see Grajczyk and Zollner, 1996).

Despite several 1980s studies that otherwise show a clear correlation between technophobia and age, Brosnan is reluctant to link the two, except to say what is clear is that the earlier individuals have their first experience with *computers*, the less anxiety they will experience with them. His account interchanges 'technology' with 'computers' at the peril of much confusion. 'As the diffusion of technology throughout many aspects of life has exposed virtually everyone to computerization, the relationship between anxiety, age and experience has become less clear' (21). The problem with this statement is that Brosnan amalgamates all technologies into one experiential maelstrom, for which he takes PCs to be a metonym. As we saw with mobile phones and SMS, the basis for generational differentiation can have a very short history and be concentrated in extremely narrow sub-media of New Media.

Network communities

The foregoing perspectives on types of bonding with both personal communication technologies and media objects deals with the more visible kinds of rituals that people have with media, via the embodied interface that they have with actual physical media. In large measure the concrete subject–object nature of this relationship provides ready evidence for ritual cases to be put concerning media. However, it may do so whilst hiding the less tangible relationship individuals may have to mediums.

Nevertheless, as we shall see, the problem with studying network communities, or 'virtual communities', as they are often called, is that they can easily be rendered as metaphysical and abstract as the 'personalized' relationships are concrete. Moreover, the various theories of virtual community, what Rheingold (1994) has called a 'bloodless technological ritual', can attain truly theological meanings as various theorists revel in the power, the totality and the unity of the universal condition which it is said to promise.

I am not therefore arguing that communication studies should not examine interaction with mediums, which is surely of paramount importance in 'post-social' society, but something of the metaphysics of such communities and of the theories about them needs to be examined.

The imagined 'universal' community

> The world is getting smaller, and [people] travel faster, countries are being brought within hours of each other instead of days, the people of those countries are getting more and more like one great family, and whether they like it or not one day they have to learn to live like one great family. (From the British wartime film *For Freedom*, 1940)

The humanist universalism that is typified in the above statement, as much as the mistaken way in which McLuhan's maxim of the global village is taken up, is one which persists today in all manner of 'computopia' narratives. It is not simply that the means of communication have speeded up, but they have a density which reduces the 'degrees of separation' between persons. As Watts (2000) claims, with the Internet, 'within a significant chunk of the www every site could be reached from every other site through about four hotlinks' (4). Thus, in Durkheimian terms, the 'dynamic density' produced by such an increase in communication infrastructure brings greater and greater numbers of persons into its orbit.

Certainly McLuhan, in his populist interpretation, has been taken up as the patron saint of such a concept of community. As Arthur Kroker (1995) says of McLuhan, he never deviated from the classical Catholic project of seeking to recover the basis for a 'new universal community' in the culture of technology: 'The Christian concept of the mystical body – all men as members of the body of Christ – this becomes technologically a fact under electronic conditions' (McLuhan in Stearn, 1968: 302). The body, both mystical and sensual, is integral to McLuhan's media cosmology, which has undoubtedly had enduring appeal among those who continue to deify the electronic village (esp. *Wired* magazine).

McLuhan's earlier advances of a techno-spiritualism find continuity in the cyber-soul or the cyber-mind in which individuality itself is resolved into a unified identity. Mark Slouka in *War of the Worlds* (1996) says:

> ... in the very near future, human beings will succeed in wiring themselves together to such an extent that individualism as we know it today ... will cease to exist. What will take its place? The great truth of our collective identity, made clear and apprehensible through the offices of that 'global mind', the Net. (96)

Such a cyber-mind is close to Gibsonian renditions of cyberspace as an out-of-body consensual hallucination which is available to be shared. A variant of this is the cybernetic unification of souls on an almost ineffable astral level whose eternal object of desire is 'community'. In her analysis of narratives of cyberspace as a gateway for spiritual redemption, Margaret

Wertheim (1999) quotes from a 1997 virtual reality conference paper by the developer of VRML (Virtual Reality Modelling Language), Mark Pesce. In opening his paper Pesce claims that those who fully appreciate what cyberspace is are those touched by revelation, sangrail, the Holy Grail – a revelation that was once the privilege of witches and mystics, which hackers have a special understanding of, and which is available to all in the modern era: 'The revelation of the Graal is always a personal and unique experience. … I know – because I have heard it countless times from many people across the world – that this moment of revelation is the common element in our experience as a community' (253).

In Pesce's account, the ecstasy of cyberspace is that it is, at once, the final paradise available to all, whilst offering all a unique, irreplaceable moment of revelation – a place of both redemption and solipsism. The excessively individualistic form of this revelation paradoxically embraces some order of a generalized other whilst failing to identify with any particular others. Foster (1997) argues that '[s]olipsism, or the extreme pre-occupation with and indulgence of one's own inclination, is potentially engendered in the technology [of the Internet]' (26). As the private basis from which each netizen realizes his or her own very personal Grail becomes more convincing, sources of self-identity become tenuous: 'as the private becomes more all-encompassing and the image of one's own world view becomes more convincing, one can lose sight of the other altogether' (26).

In the absence of the other, avatars everywhere encounter only themselves, and extend themselves in the image of the medium itself, which acts back on them as an externalized spirit to be worshipped. Wertheim (1999) remarks that, '[i]n one form or another, a "religious" attitude has been voiced by almost all the leading champions of cyberspace' (255, see also Robins, 1995: 151). Regardless of whether the champions of cyberspace are 'formal religious believers' like *Wired* magazine's Kevin Kelly, 'again and again we find in their discussions of the digital domain a "religious valorization" of this realm' (256).

In order that such a place is reserved for cyberspace, it is frequently associated with Judaeo-Christian narratives of which it is seen to be part of an eternal return. For example, Michael Benedikt (1991) describes it as a New Jerusalem, which, like the Garden of Eden, 'stands for our state of innocence' and is a 'Heavenly City which stands for our state of Wisdom and Knowledge' (15). In the Book of Revelations, the Heavenly City exhibits a beauty and a geometry tantamount to a 'religious vision of cyberspace itself' (Wertheim, 1999: 258).

However, the genesis myths which are commonplace are not at all restricted to Christianity.[9] As Wertheim points out, they can just as easily be based in Greek mythology: 'From both our Greek and our Judeo-Christian heritage Western culture has within it a deep current of dualism that has *always* associated immateriality with spirituality' (256).

The ineffable and the sublime feature strongly in Greek mythology, and indeed the very idea of utopia comes from the Greek, meaning 'nowhere', a beyond which cannot be found in the mundane materiality of everyday life.[10]

Of course, technology has always figured in attaining such a beyond, for which cyberspace is the latest enlistee. Because of this, the Greek legends used to explain technologically mediated community are always about the emancipation as well as the imprisonment which technology is capable of delivering. In *Technopoly* (1993), Postman narrates the Judgement of Thamus as being able to teach us that 'every culture must negotiate with technology', and that always 'a bargain is struck in which technology giveth and technology taketh away' (5). The realization by human beings of some form of community is said to be dependent on the correct negotiation of the perils and ecstasies of technology.[11]

These relationships to technology invariably harbour dreams of *unity*, in the form of democracy, community and a metaphysical common weal or common interest. In nearly all cases the status of modern communication technology is cast within the above configuration of a 'use or abuse' framework of technology. Technology can be a source of great enlightenment, hope and belongingness, but it is assumed that it can also foreclose relationships in which it has a lesser mediating role.

Such is the tenor of Darin Barney's *Prometheus Wired: The Hope of Democracy in the Age of Network Technology* (2000). Barney takes a rather more in-depth look at the history of both technology and the meanings of political utopia, employing the writings of Plato, Aristotle, Marx, Heidegger and George Grant for this purpose.

In doing so, Barney revisits the myth of Prometheus, who concealed the earthly possession of fire from Zeus. Zeus's response was to deny the blessing of hope to mortal beings. The technology of fire is granted. It may provide light and warmth, but Zeus knew it would also lead transient beings to extend themselves beyond their mortal limits, leading to apocalyptic situations. 'When beings who are mortal by nature no longer foresee their own death, they begin to regard themselves as *im*mortal: as having no natural limits, like gods, which they are not. Hope thus seduces human beings into overestimating and overreaching themselves, with tragic consequences' (5). Barney's corrective to Enlightenment doctrine is that '[h]ope enlightens, but it also blinds' (5). Nowhere are the stakes of this insight so high as they are in the virtual dreams that are held out for the role of network technology in political organization.

The immateriality of cyberspace, the fact that it is so 'clean' and free from mortal struggle, coincides with another metaphysical theme of the cyber-utopians – that of 'liberation from the flesh'. As Wertheim (1999) has commented, 'Among many champions of cyberspace we also find a yearning for transcendence over the limitations of the body' (259). In the sentiments of cyber-enthusiasts like Jaron Lanier and Nicole Stenger, it is possible to 'transcend the body' or at least be 're-sourced' sparkling juveniles who will never age (259). 'Dreaming of a day when we will be able to download ourselves into computers, Stenger has imagined that in cyberspace, we will create virtual doppelgangers who will remain youthful and gorgeous forever' (259).

Given that the virtual community can never be 'lived' in embodied form, but only imaginatively by extension, it is not surprising that the imagined universal community is also realized in a secular quest for liberation from the flesh itself. The intellect, rather than the body, lies at the centre of the life world of the virtual community, where 'the secular and scientific myth of the conquest of nature could be fused with the logic of commodity production' (Sharp, 1985: 65).

Abstraction from the flesh (solipsism) and spiritual unity (universal humanism) are at the base of a bewildering array of attempts to define a kind of *'cybergeist'* in some form of conceptual singularity. Solipsistic tendencies to *define-it-ism* are a telling indication of this. Typically this entails inventing neologisms which have the minimal reference to comparable terms whilst claiming the maximal reference to a universal condition. The authors of these neologisms stand out as ideologists of this solipsism, which can attain rapturous proportions. Numerous, zine-like publications flourished in the mid-1990s advancing home-grown concepts of cyberspace which made little attempt to engage with concepts which already addressed their 'investigations'.

We have already discussed the *Apparatgeist*, but numerous other examples can be given. Sometimes terms might be referring to the same thing but talk past each other as they each rush in to claim a universal descriptor that is oblivious to the rush of neologisms found elsewhere. *Apparatgeist*, digitopia, cyberutopia, cyberia (Rushkoff), technopoly (Postman), infomedia (Koelsch): lexical fragmentation prevails at the precise juncture where an homogeneous speech community is supposed to find renewal.

One even finds completely needless attempts to redefine those few terms which *have* established themselves – such as 'cyberspace'. An indulgent case of this can be found in Darren Tofts' *Memory Trade: A Prehistory of Cyberculture* (1997), an illustrated essay which feels compelled to redefine cyberspace as 'cspace'. It is worth quoting at length:

> To come to terms with the historicity of cyberculture we need a concept that identifies both the ur-foundation of technologized consciousness, as well as its extension in the current preoccupation with the creation of digital worlds. The concept I propose is called 'cspace'.
>
> The concept of 'cspace' first announced itself as a means of abbreviating cyberspace, a nonce invention that served the purpose of expedience. I first used the term when I was thinking through the connections between poststructuralism, cybernetics and writing. Within that context, it took on a new, aleatoric meaning, embodying many of the ideas I was working with at the time. Pronounced in exactly the same way as 'space', cspace is beautifully ambivalent, for as a form of shorthand it accommodates two different meanings (cyberspace/space), yet they cannot co-exist

at one and the same time. A particular kind of reading must choose between them ... no spoken reading of cspace can elide cyberspace with space. Furthermore, a silent reading of the text on the page must unpack and mentally vocalize cyberspace, codify the visual sign with acoustic value, in order to hear it. (51)[12]

For Tofts, 'cspace', which is initially announced as a concept, rapidly progresses to an ontology which has 'manifestations' such as the everyday division between 'inside' and 'outside' (51) of technological worlds, whether this be 'jacking into the matrix' or 'speaking on the telephone'. Cspace is heralded as a kind of ancestral trope, the inside and outside of alphabetic writing, which is said to invoke Socratic fears about language having spatial architectures that cannot be controlled by the subject. For this reason 'cspace', it is repeated, is no less than the 'ur-concept of technologized consciousness, as well as the grammacentrism of cyberculture' (51).

A very different but equally inflated conflation of a nomenclature with an ontology is that of 'cyberpower', put forward by Tim Jordan (1999). Writing at the end of the 1990s, Jordan felt confident enough to proclaim:

The patterns of a virtual life are clear enough to be mapped. The virtual world and its social order can be traced now in its entirety, from pole to pole. This does not mean all areas are perfectly known. Sometime in the future we will probably look back at this map and see where it has equivalents to the dragons and the sea monsters faithfully represented on early maps of the world. (3)

This nevertheless does not prevent a self-assured vision of cyberspace as a totality:

However, we can produce an overview of all of cyberspace's multifarious life, the first globe of cyberspace. This book is such a globe. It is a cartography of the powers that circulate through virtual lives, a chart of the forces that pattern the politics, technology and culture of virtual societies. These powers set the basic conditions of virtual lives. They are the powers of cyberspace and together they constitute cyberpower. (3)

As the book proceeds, cyberpower turns from being a map of power relationships on the Internet, to being indistinguishable itself from the form of this relationship: 'Cyberpower is the form of power that structures culture and politics in cyberspace and on the Internet' (208). The merit of Jordan's book is that it does break down the forms of recognizable power relationship that operate in cyberspace in useful ways: individual, social and imaginary. But as with Tofts' fixation with his own solipsistically invented 'monster concept' of cspace, Jordan valorizes cyberpower into a theoretical dragon too indeterminable to have any analytic value.

The virtual Internet community

It can be seen from the foregoing discussion that speculation on network communities can attain quite theological dimensions. At precisely the time that Alain Touraine made his sociological claim that social systems, as systems, no longer exhibited transcendent 'universal unifying values', some cyber-utopians wanted to point to the Internet as providing just such values. However, this is not true of a growing body of empirical and analytic research which has been conducted from the mid-1990s onwards and which has sought to unravel the specific forms of connection and bond which the Internet makes possible. Such research can be divided into two clear methodologies and premises. One body of research merely extends demographic sociology to include questions about Internet use (Anderson and Tracey, 2001; Di Maggio et al., 2001, Howard et al., 2001; Nie and Erdring, 2000; Wellman, 1999), whereas the other form of research focuses exclusively on Net communities in their various forms (e.g. Baym, 1998, 2000; Rafaeli and Sudweeks, 1997; Smith, 1997). The demographic researchers proceed much more from a behaviourist 'impacts' paradigm, asking questions like 'are on-line identities consistent with off-line identities?', whereas the virtual community studies are interested in the *sui generis* qualities of the new medium, exploring whether a new medium allows for new ways of behaving and new identities which bear no relation to, or cannot be meaningfully compared to, off-line identity.

For these latter researchers, 'virtual' does not mean immaterial and spiritual. Virtual communication might be disembodied, but it has a definite architecture and technical infrastructure which is material – a network rather than a matrix.

Rather than run headlong into announcing a mythical universalism, the empirical research on Net communities has taken the Internet as a model for specifying a set of social dynamics which can be distinguished from either 'broadcast' (as the first media age) or face-to-face communities.

Certainly, it is easy to differentiate 'virtual Internet communities' from face-to-face communities. They are, as in Dempsey's second type of community discussed above, made up of persons who do not necessarily know one another but have a sense of belonging together (Foster, 1997: 24). And the fact that such communities are entirely disembodied in no way lessens the solidarity of such a community in the Durkheimian sense.

Despite the fact that audience communities are able to constitute a mediated ceremony, as we shall see, the term 'virtual community' has only attached itself to the Internet. Clearly, audience communities qualify as 'disembodied communities' in which persons still feel a sense of 'belonging together'. We saw, however, in Table 5.3 in the previous chapter how this feeling of 'belonging together' is mediated by different agents (technical infrastructure: Internet; and technical plus human infrastructures: broadcast). Both Internet and broadcast also exhibit some of the qualities of face-to-face communication, but in inverse relationship: interaction without reciprocity (Internet) and reciprocity without interaction (broadcast).

What is valorized in Internet communities is that they are 'interactive' (Willson, 1997: 146). Interactivity becomes foundational for the speculation of virtual community and, as particularly the second media age theorists argue, for community in general. We are reminded by Rheingold (1994) that virtual communities are a compensation for the loss of traditional communities around the world. The loss of these communities has been postulated by Stone (1991) as a 'drive for sociality' – 'a drive that can be frequently thwarted by the geographical and cultural reality of cities' (111).

What is denied in the metropolis is communities of a 'face-to-face' nature, and their attenuation is seen to be compensated for by the electropolis, which, whilst distinguishable from these communities, is largely thought within their terms: 'More and more commercial, political, media, and social interaction occurs in cyberspace every day, supplementing and enhancing face-to-face meetings. A new class of acquaintance has emerged: virtual friends and associates – those we feel we know well but have never met' (Whittle, 1996: 230).

Because 'virtual Internet communities' are delineated within the face-to-face paradigm, they work to emphasize every quality of on-line communication which is intrinsic to face-to-face communication whilst explaining those qualities that are absent either as a loss, or as compensated for in other ways. What is retained by the utopian theorists of virtual Internet communities is an homogeneous sense of an *agora* in which interaction can take place, and, of course, interactivity itself, which is raised to an ideology of the modern period.[13]

Rheingold (1994) is an early paradigm example of this view. For him, cyberspace has the potential to be a place which 'can rebuild the aspects of community that were lost when the malt shop became the mall' (24–6), named variously as the new 'social commons' or the 'electronic *agorae*'. The fact that an *agora* that is global in proportion can replace the malt shop does not present problems for Rheingold. This is because an *agora* is only limited by its *exclusion* of interactivity. If the virtual community can facilitate the latter, then, no matter how many participants, it still qualifies as an *agora*.

Of course, for second media age figures like Rheingold, the actual *agorae* of pre-media society are scantily theorized. What are projected and idealized are those geographical communities, antedating broadcast and the culture industry, that managed to thrive in urban life and rural centres in various forms of 'third places' – the market, the promenade, the street and the arcade. But the confluence of mass media with the 'automobile-centric, suburban, fast-food, shopping mall way of life eliminated many of these "third places" from traditional towns and cities around the world' (Rheingold, 1994: 25).

For other second media age theorists, like Mark Poster, the *agora*, a foundation of ancient Athenian civic life, is deemed distinctive in its openness, unregulated by the state or church, and an arena for unfettered political expression. With the growth of cosmopolitan cities in early European modernity, the *agora* typically contracted to institutions which became normative for embodied interaction – the cosmopolitan coffee house predominates as the most significant continuation of a public *agora*. But also, as Poster (1997) points out, 'the New England town hall, the public square, a convenient barn, a union hall, a park, factory lunchroom, and even a street corner' may perform *agora*-like functions (217).

Roseanne Stone (1991) takes up the virtual-community-as-agora thesis so sweepingly that she constructs an entire genealogy of virtual community, including its mythologies, that dates from well before the Internet itself.

As Ostwald (1997) outlines it, Stone's 'phenomenological view of the spatial and the experiential' is divided into four epochs of virtual communities.

The first epoch brings together intellectual interchange, which survives today in the university: 'the academic community of the journal has, like any other community, strict laws and customs. Communications between members of the community may be rigidly ordered to meet accepted forms of language, referencing and format' (131).

The next epoch of virtual communities derives from mass media: 'In this epoch, the virtual spatiality of radio and television connect people together through perceived experiences and the illusion of participation. Like the academic community spatialized within the journal or paper, the tele-visual community creates its own particular variations of language and presentation' (131).

The founding moment for the third epoch was when the intranet network, Communitree, went on-line in May 1978. An asynchronous bulletin board capable of facilitating CMC, 'Communitree was a rudimentary precursor to the global news-nets where hundreds of thousands converse daily and exchange data in a free-flowing system' (131).

Epoch four in Stone's virtual community is based on Gibson's 'Matrix', a form of cyberspace in which the communion of selves attains its fullest expression. The mere existence of multiple selves in cyberspace (the socialization of virtual reality) guarantees an interactive freedom unparalleled in the other epochs.

Stone's epochs provide a somewhat more nuanced way of thinking about modes of communicative association. But what is common to all of these epochs is that *agorae* of interactivity are their recurring basis.

Whilst Stone distinguishes between different *agorae*, and describes the kinds of community they make possible, the nature of *interaction* which takes place within them is left untheorized. As in the case of second media age thinkers, it is often supposed that interaction is a matter of speech, or, at least in the Gibsonian Matrix, an exchange of consciousness.

But does community always require a reciprocity and the exchange of consciousness? This view is certainly a pervasive one, for which the *agora* provides a gathering-as-communion, in which interactivity is always modelled on face-to-face exchange.

The sense of community made possible by flânerie

The practice of *flânerie* – a form of pedestrianism which sought the sensation of crowds for its own sake – is, as it has appeared at different times in the development of cities, I argue, indispensable for understanding virtual Internet communities, which can be viewed as a continuation of *flânerie* by other means.

We have seen already the difference between *Gemeinschaft* and *Gesellschaft* inherited from classical sociology. As narrated by Tönnies, this difference is borne out by the contrast between city and village. The village is itself an *agora* replete throughout its entire aspect, whereas in the city certain places are set aside especially for the conduct of interaction. As is prevalent in much of the literature, the virtual community takes the village, or the 'global village', as its analogue, with a selective interpretation of McLuhan providing a ready-made justification. In the next two sections, I am going to argue that cyberspace does not signal the return of the communitarian *agora*, but is in fact a continuation of the *cosmopolitan agora*. Such an *agora* is not distinctly European but can be found in all cultures which undergo rapid urbanization. It has more to do with the culture of cities than with culture ethnically defined; indeed, this is what makes the *flâneur* such a precursor of a global form of cosmopolitan identity.

But to understand this, we must understand the changing practices of *flânerie* which emerged in European modernity. Up until the mid-nineteenth century in Europe and Russia the public promenades of the great cities drew massive crowds, where individuals began a practice that is scarcely visible today, but which has been heralded as a defining practice of modernity – that of the pedestrian who seeks out the crowd. Two principal *savants* of *flânerie* are Charles Baudelaire (1972) and Walter Benjamin (1977), who describe how, at the height of the period of *flânerie*, *flâneurs* revelled in their anonymity. Published in 1863, Baudelaire's benchmark essay 'The Painter of Modern Life' declares of the *flâneur*: 'The crowd is his domain, just as the air is the bird's, and water that of the fish. His passion and his profession is to merge with the crowd' (Baudelaire, 1972: 399). In the rapidly populating cities, in the tumult of the revolutionary world, the individual becomes at once an outsider and singular, but, as Featherstone (1998) notes, there is a recognizable social type: 'the flaneur moved through the crowds with a high sense of invisibility – he was in effect masked and enjoyed the masquerade of being incognito' (913).[14] The kind of space which the cosmopolitan *agora* offered to the *flâneur* was not one which redeemed *Gemeinschaft* in the midst of the multitudes. Nor did it produce anomie either (Tester, 1996: 7). Rather, for the bourgeois stroller, it was a place of intense fervour and passion.[15]

The promenades of the great cities of modernity were places of high-volume interaction, but very impersonal interaction. The main feature of such 'face-to-face' interaction was that it was *visual*.[16] It was either to see or be seen that the *flâneur* sought out the crowd. With such purpose the *flâneur* would overcome unfeeling isolation through enjoying the multiplication of selves. The metropolitan person is also addicted to the heightened tempo of the large city – and would form an attachment to the street as if it were a home.[17]

For Benjamin (1977):

> The street becomes a dwelling for the *flâneur*; he is as much at home among the façades of houses as a citizen is in his four walls. To him the shiny, enameled signs of businesses are at least as good a wall ornament as an oil painting is to the bourgeois in his salon. The walls are the desk against which he presses his notebooks; news-stands are his libraries and the terraces of cafés are the balconies from which he looks down on his household after his work is done. (37)

As Keith Tester (1996: 5) explains, the *flâneur* has a calling, which is *doing*, not simply being. For Baudelaire (1972), the man who lives in a box, or the man who lives like a mollusc (the man who simply is), is actually incomplete; the struggle for existential completion and satisfaction requires relentless bathing in the multitude (it requires doing over and over again).

The *flâneur*, then, 'the man of the crowd', works at his identity, and does not actually lose it in the *crowd*, as the *badaud*, or simple 'man *in* the crowd', does. His observation of the street is specialized and intellectual. The aestheticization of such a gaze provides the *flâneur* with a sense of differentiation; it is what makes him special, and individual, even where in fact he may be repressing the images of decay which can also be observed around the city.

For this reason, the *flâneur* lives heroically as a foreigner in his own city[18] – the complete reverse of the situation of the village or small town person. For the latter, strangers only come from outside and must be assimilated – as Shields (1994) suggests: 'the stranger is a foreigner who becomes like a native, whereas the flaneur is a native who becomes like a foreigner' (68).

If there was community in the town square and the cafés of mid-modernity it was in the display of *proto-cosmopolitanism*. But in both forms of the *flâneur* it was interacting with a mass of strangers which became life's prime goal: to be outside, to be with others for the sake of it, *not to anticipate an individual interaction*. As we shall see, the form of individual that arose from *flânerie*, the person who swung between being seen and being invisible, is very important for understanding the Internet avatar, whose community is also to mix with those who are strangers. Like the *flâneur*, the avatar is invisible, but mixes with the broadest kind of 'public' ever envisaged, that of the virtual community.

From street to virtual flâneur *– the transformation of* flânerie

Click, click, through cyberspace, this is the new architectural promenade. (Mitchell, 1996: 24)

For Walter Benjamin, *flânerie* was a peculiarly nineteenth-century European phenomenon, whose fall from public life resulted from the withdrawal of spaces in which it could be practised. The street itself was undergoing massive changes. Pedestrianism of any kind had become hazardous as the tramcars, the trolleys and locomotion drove corridors of speed throught the cities. As Benjamin narrated it, this is why the great arcades had become so important, as they turned the street inward and quarantined an exclusive zone in which the *flâneur* could thrive (Benjamin, 1977: 36). But when the arcades began to be demolished to make way for department stores, it was the beginning of the end of this kind of *flâneur*.

If the arcade is the classical form of the *intérieur*, which is how the *flâneur* sees the street, the department store is the form of the *intérieur*'s decay. The bazaar is the last hangout of the *flâneur*. ... if in the beginning the street had become an *intérieur* for him, now this *intérieur* turned into a street, and he roamed through the labyrinth of merchandise as he had once roamed through the labyrinth of the city ... (Benjamin, 1977: 54)

The story of the disappearance of the *flâneur* coincides with the story of the disappearance of public space. The street becomes the arcade, the arcade becomes the department store, the department store is absorbed by vast privately owned zones that are shopping malls.

For Mike Featherstone (1998: 910), in the modern period this means that urban spaces where the occupants of different residential areas could meet face-to-face, engage in casual encounters, accost and challenge one another, talk, quarrel, argue or agree, lifting their private problems to the level of public issues and making public issues into matters of private concern – those 'private/public' *agorae* of Cornelius Castoriadis are fast shrinking in size and number.

For Benjamin, this process corresponds to a change in the *flâneur* also, who is redefined as a consumer, a private, possessive individual who attempts to re-create his own sense of a world-of-the-whole through the market. The final phantasmagoria is the private dwelling itself, which, as Ann Friedberg (1993) describes it, simulates a dioramic display of goods and commodities, and, together with media, provides for every need.

As we saw in the previous chapter on mobile privatization, the private home itself becomes the basis for a virtual *agora*. Electronic assemblies and electronic interaction take precedence over interaction with our neighbours, or geographic community. Indeed, as McLuhan says of the United States at least: 'to go outside is to be alone' (quoted from an interview with Tom Wolfe, 'McLuhan: The Man and His Message'. Staying inside becomes the basis for connection. There is no longer any calling to *physically* associate as a goal in itself.

In exploring this de-physicalization of *flânerie*, Featherstone (1998) asks, 'How far can the new forms of electronic communication such as the Internet and multimedia also facilitate *flânerie*?' (919). A number of commentators (including Featherstone, 1998; Jones, 1995; Mitchell, 1996; Smith, 1999) have insisted that cyberspace is a unique replacement for the geographic agora. As Steven Jones has suggested of CMC,

> Like the boulevardiers or the denizens of Nevsky Prospect described by Berman (1982), the citizens of cyberspace (or the 'net' as it is commonly called by its evanescent residents) come here to see and be seen, and to communicate their visions to one another, not for any ulterior purpose, without greed or competition, but as an end in itself. (Berman, 1982: 196, cited in Jones, 1995: 17)

For them, broadcast cannot offer a meeting place of this kind. Instead, it gathers audiences together but leaves them with little to discuss save for what is offered as spectacle. Indeed, the sub-media of the Internet are able to offer the same kinds of opportunity for interaction as did the spaces of *flânerie* which featured in the early period of modernity.

For Featherstone (1998), the most important characteristic of Internet sub-media in this regard is that they are *non-linear* – they move away from 'the linear physical construction of the book with its sequence of pages, or the film with its one-way movement through time' (9). Such forms of association were not possible with television until the development of video cassette recorders/players. But the database, whether it is a CD-ROM or the Internet's 'Library of Babel', enables instantaneous information association, which, according to Featherstone, encourages electronic *flânerie*. But unlike the deliberate ambling of the traditional *flâneur*, hypertext allows the virtual traveller to jump to other places in texts or in the Web of pages. Such *flânerie* presupposes the ontology of the disembodied-extended, discussed in the previous chapter, the individual who is 'lifted out' of the constraints of embodiment, or, as Featherstone puts it, does not have to 'wait to reach the street-corner to change direction'.

> Indeed, the jump, to continue the metaphor, can be to another city. Not only is the *flâneur*'s city a world, but the world has become his/her city: with everything potentially accessible, potentially visible. Hence, hypertexting brings to the fore the problem of navigation, of movement within text or work. There no longer is a correct sequential way of reading or proceeding; the work has become directly activated by the particular purposes, or whims, of the user. This is one of the defining characteristics of the new electronic media: interactivity. It encourages engagement and two-way interaction on the part of the user which contrasts to the one-way mode of communication, which encouraged passive reception, which we find in the traditional age of mass media. (921)

So, once again, we find a central characteristic of the second media age being introduced here as an agent of the return of *flânerie*. Electronic *flânerie* is also an improvement, in speed, mobility and reach, on the urban *flâneur*.

> The urban *flâneur* typically sauntered around, letting the impressions of the city soak into his subconscious. The electronic *flâneur* is capable of great mobility; his pace is not limited to the human body's capacity for locomotion – rather, with the electronic media of a networked world, instantaneous connections are possible which render physical spatial differences irrelevant. (921)[19]

Featherstone does, however, tend to reduce such non-linearity to 'the Internet' as an exclusive environment for such mobilities. For example, he argues that 'with the Internet there has been a massive speed up of the rate at which new perceptions are brought in front of the eye' (921–2), whereas, as we know from Simmel (1971), the 'mental life' of the metropolitan person is before all else distinguished by 'information overload'. The second media age thesis dilutes also when we compare the largely textual experience of the 'World Wide Wait' with the frenetic culture of music television, which institutionalizes channel surfing, or image surfing, into a genre, as Featherstone himself observes: 'we can recall the much-vaunted postmodern channel hopper, MTV, or music video viewer, who is bombarded with fragments of images and information removed from their context so that he is incapable of chaining together the signifiers into a meaningful message' (922). This quote would insinuate that the mass media do not generate such passive participants as Featherstone earlier suggests.

An added counterpoint of Featherstone's second media age argument is his assertion that virtual forms of association preceded the Internet by many years. The difference between today's 'city of bits' and the industrial city is one of degree only. The industrial city was always 'an information city in the sense that the urban landscape was continually inscribed and reinscribed with information, with cultural meanings in the aestheticised façades of buildings, advertisements, neon signs, billboards' (922). And we could add to this observation the fact that nineteenth-century technologies of simulation were already providing alternatives to the mutual *agora*. The panorama, the diorama and cinema prefigure television and the Internet by providing virtual spaces to 'travel' which give the illusion of being part of worlds that do not require physical involvement.

Problems with dominant definitions of virtual community

The fact that cyberspace may provide a milieu for electronic *flânerie* conflicts with two very prevalent views of virtual Internet community. The first rests on its touted efficiency in being able to connect people with common interests,[20] and the second is that it is able to reduce serendipitous interaction between strangers.[21] Both of these views fall within an instrumental conception of the Internet.

In the first case, virtual community is cast in terms of interest-based relationships.[22] In the community-of-interest view, individuals, or avatars, are bonded together solely in the pursuit of common interests. See Marc A. Smith for a comparison of virtual communities with the committees of correspondence of the eighteenth and nineteenth centuries. The appeal of CMC or the Internet is that it is seen to deliver the satisfaction of these interests much more efficiently than can life off-line. However, the needs that are met to satisfy these interests tend to be less material than they are psychological. And indeed the appreciation of *shared* interest in a particular movement, style or object, be it the environment, fashion or tropical fish, is viewed to be sufficient in most cases for a sense of bond to emerge with other people. Of course, this particular view, heavily promoted by *Wired* magazine in the 1990s, is founded in a consumerist approach to community. J. Macgregor Wise (1997) argues that it didn't take long for the Internet to become a frontier for commodification, not in what it sells, but in the way individuals relate to it: 'Community based on special interests (hobbies) is already on its way towards a consumerist-centered (rather than community based on community interests) organization, where the dominant communities (e.g. newsgroups) are communities centered around leisure activities (e.g. *Star Trek*)' (154).

Another dominant instrumental view of the virtual community is celebratory of on-line cultures where there is 'no sense of place'. In these definitions, computerization removes dependence on physical 'place' and minimizes accidental contact amongst strangers. This in turn provides an architecture for the community-of-interest to flourish, as individual interactions are unlikely to be cluttered by colliding with unwanted forms of association. Thus, it enhances the efficiency of communication among those who already need to communicate more efficiently (see, e.g., Calhoun, 1986). However, this latter view differs substantially from the practice of *flânerie*, which encourages a community of strangers bathing in each other's company.

In both of these views, the idea of community is a *Gemeinschaft*-by-other-means, and not association as an end in itself. In both views, traditional 'nostalgic', real-life senses of community can benefit, but they can also be threatened. Wellman and Gulia (1999) see the utopian versus the dystopian versions of virtual community arguments as the central binary which organizes the debate.[23] But the binary between the reinvigoration of real-life community and its attenuation conceals the fact that what is reaffirmed by both positions is a commitment to a one-dimensional sense of real-life community, which can be either 'helped' or eroded by technological mediation.

One glaring problem with such a commitment is that, as we have identified, it is historically possible to uncover senses of community not anchored in communicative interaction, but which *also* do not involve technological mediation – a fact which problematizes too simple a real-life/virtual community distinction.[24] The merit of the *flânerie* perspective, for example, is that it points to long-held practices of association which have very different bases than that of functional connections between rational, autonomous subjects. It allows that the visual, aesthetic and mutual sensoria of public life are also strong sources of community feeling. By contrast, the widely heralded instrumental functions of cyberspace are only a part of the spectrum of human interaction. Moreover, the hailing of the Internet as a redemption of interactivity accentuates the way CMC dramatically extends dialogic communication as an instrumental aspect of face-to-face communication, but at the expense of other qualities of mutual presence.

The more reliant individuals are on CMC to meet communication needs, the less they engage with embodied interaction. No empirical studies are necessary to demonstrate this relationship.[25] This, in itself, is not a dystopian situation; it is just that it precludes the possibility of more rounded (multi-levelled) expressions of everyday life. This is usually expressed by critics of on-line communication via a broader critique of computerization cast within a humanist narrative. George Lakoff (1995) laments:

> One of the sad things is that the increase in computer technology does not get you out into the world more, into nature, into the community, dancing, singing, and so on. In fact, as the technology expands, there is more expectation that you will spend more of your life on a screen. That is not, for my money, the way one should live one's life. The more that the use of computers is demanded of us, the more we shall be taken away from truly deep human experiences. That does not mean you spend time at a computer, you will never have any deep human experiences. It just means that current developments tend to put pressure on people to live less humane lives. (124)

Once again, the entrenched dichotomy which opposes a 'deep' human essence to technology appears here. Rather than on-line interaction being merely a new kind of human experience which may or may not result in a loss of other levels of human experience, it is seen to result in a decline in 'human experience' in general.

Lakoff's othering of technology is also evident in his claim that experience on-line is fictive:

> Online 'interactivity' is an illusion. What passes as interactive is pretty uninteractive! It has to do with some fixed menu, not with being able to probe as you would a person or to judge or be moved as you would in a live interaction. There have to be canned answers and canned possibilities. (Lakoff, 1995: 24)

We have also seen that community has many possible sources, and that cosmopolitan settings also make a form of community possible via the image and the effervescence of large populations. Echoing Lakoff's charge that the interactivity of on-line communities is not genuine is the concept of 'pseudo-communities' (see Beninger, 1987; Jones, 1995; Slouka, 1996; Stoll, 1995).[26]

What is missing in the critique of virtual community as 'pseudo-community' is a basic but important understanding of the fact that 'virtual community' can seldom be separated from physical community. Too often, attempts are made to conduct isolated analyses of Net interaction in isolation from other interactions without considering that '[t]he Net is only one of many ways in which the same people may interact' (Wellman and Gulia, 1999: 170). The empirical research of Wellman and Gulia and the analyses of Castells (2001) are particularly concerned to dispel the dichotomy between an idealized past of parochial community and the narcissistic future of physical isolation overcome by virtual association. Castells wants to show that uses of the Internet are 'overwhelmingly instrumental' and 'closely connected to the work, family and everyday life of Internet users' (118). He is at pains to demonstrate that on-line activity does not detract from off-line life, and selectively reviews surveys which show 'that internet users have larger social networks than non-users' (121). Of course, what Castells reinstates is precisely the dichotomy between real/virtual that he says he wants to overcome. In repeating what is a commonplace habit in conceptualizing virtual communication, Castells does not consider use of the Internet to itself be *already* part of 'everyday life' or already 'social'; rather, it is seen to be simply compatible with, or to enhance, such life and not be a threat to it. In arguing that email, for example, is predominantly used for 'work purposes, to [do] specific tasks, and to keep in touch with family and friends in real life' (118), Castells' actual object is not virtual community or the Internet, but institutional and face-to-face 'real life', defined as 'real', in relation to which communication technology provides a service. Such an approach is incapable of anticipating precisely what is *sui generis* about virtual community, or that new forms of electronic assembly may exist which enable forms of interaction not previously practised. To focus on such novelties of interaction is not to inflate them to the status of a *Zeitgeist*, but to establish methodologies to analyse the nature of a new kind of social bond.

Overcoming this recurring dualism between real and virtual, Kumiko Aoki makes a useful but regularly overlooked distinction between three domains of virtual community:

1 those which totally overlap with physical communities
2 those that overlap with these 'real-life' communities to some degree
3 those that are totally separated from physical communities (cited in Foster, 1997: 24).

In his review of empirical research discussed above, Castells (2001) claims that most Net interactions fall into category 1 or 2. Based on this research, he concludes that most accounts of virtual community are over-inflated, and that this has led to a simplistic stereotyping of most Net users as isolated nerds and geeks who live out a withdrawn and isolated existence. As I have argued above, Castells' claims are based on the collapsing of demographic surveys which set out from an institutional framework with research into CMC. This does not leave him with any room for a considered analysis of category 3. Either such a category is reserved for a tiny minority of persons with identity problems who seek the Net as some sort of fantasy or refuge from real life, or else Castells is anxious to show that such persons also have a life off-line where they lead quite normal existences, and are 'civically engaged'; i.e. that there is no withdrawal from other forms of sociability. This again shows that, for Castells, the 'social' is a decisively off-line condition and that life on-line either enhances or annuls such sociability.

It is true that, where there is total overlap between the two kinds of communities, we typically think of on-line communication as an augmentation or supplementation of face-to-face community rather than the other way around. In the case of partial overlap, the moment of *identification* between two communicants is often more ambiguous and difficult to negotiate. For virtual communities which have no reference to physical communities, identification is not burdened with 'verification' in that identity formation is purely internal to the speech events.

Because of this, there is a sense in which, in the third situation, part of every communication event consists, in part, of interlocutors' conversations with themselves. Taylor argues that in CMC, 'The striving subject enters into conversation in order to build itself up through the search for truth. Thus the person who converses relates to herself/himself even when s/he seems to be relating to an other' (cited in Foster, 1997: 26). The computer terminal is less a window onto other worlds than it is a half-reflective mirror. Psychoanalysts would tell us that this is true of all communication, but in cases where such communication becomes technologically embedded, it also gains a different kind of visibility. Where there are low levels of identification, the case of Aoki's third form of community, the validation of truth, morality and aesthetics, the three normative domains inherited from modernity cannot be satisfied dialogically, except in an imaginary sense, an illusion that is proportional to the accountability of each communicant.

But the low accountability of on-line communicants, what I have called 'interaction without reciprocity', isn't merely related to the absence of identification between interlocutors but also to their high virtual *motility* – the potential for mobility, including the mobility to 'switch off' and escape from any communication event. This feature is accentuated by the high number of possible connections which can be made on-line, and the fact that such connections are part of a web or a network.

Rafaeli and Sudweeks (1997) have argued that on-line interactivity needs to be thought of across an entire network, not simply between two given interlocutors. For them, connectivity is more important than interactivity, and the degree to which a given interaction transcends simple reaction is decisive here.

As Tanjev Schultz (2000) summarizes their view:

> Eventually, in two-way, or reactive, communication one side responds to the other, but such communication remains reactive unless 'later messages in any sequence take into account not just messages preceded [sic] them, but also the manner in which previous messages were reactive' (Rafaeli and Sudweeks, 1997). But Rafaeli also draws a very fine line between two-way and reactive communication: 'Two-way communication is present as soon as messages flow bilaterally. Reactive settings require, in addition, that later messages refer to (or cohere with) earlier ones' (Rafaeli, 1988: 119). Rafaeli's model suggests that a lot of use of the new technologies is far from interactive. (210)

As Schultz suggests, two-way communication does not, in itself, guarantee interactivity. And to extend Taylor's point, if an exchange does not develop into a relationship where one utterance becomes a context for another, the discourse becomes egological. Conversely, reactive communication is not just typical of broadcast communication, but is possible within networks.

As we shall see in the next section, it is fanciful to see 'interactivity' as a precondition of virtual kinds of community. This is related to a final problem that is endemic to dominant definitions of virtual community, which is that it is *confined to network forms of community*. When instrumental views of virtual community are critiqued, it becomes clear that broadcast architectures also enable such communities. This further dilutes the historical distinction between first and second media age, although, as we shall see, the two forms of virtual community have their own specific dynamics.

Broadcast communities

> Virtual community is people all over the world gathered around television sets to watch the Super Bowl or a world cup match. (Wilbur, 1997: 14)

A primary implication of the contrast between broadcast and network communication discussed in the previous two chapters is the way in which the study of communication architectures allows us to reconsider broadcast as a technical medium of social integration just as computer-constituted communication networks are today viewed as a medium of identity and community. As we saw in Chapter 2, broadcast mediums are not simply conduits for messages, but facilitate institutionalized spectacle as well as constituting enclosed worlds of representation which may become self-referential. These two aspects – the spectacle and simulacrum properties of broadcast media – are active in constituting worlds of representation which attain ritual status for media audiences, and indeed become sources by which audiences acquire the symbolic and semiotic materials with which to construct a meaningful identity and a world-space of personal object meanings. As Silverstone (1999: 99) observes, the role of ritual community is confined to broadcast.

In a capitalist society these ritual processes are largely of benefit to the culture industry. But spectacle and simulacrum also generate apparently resistant rituals, which, on the surface, seem to be a reaction to the very structure of broadcast itself, but which the culture industry continuously re-appropriates.

As we saw in Chapter 3, this system of cultural production where the few produce culture for the many is widely condemned by the different celebrants of the Internet. For them, the major feature of the Internet is that it enables democratized participation in the public sphere, which in the era of broadcast and 'systematically distorted communication' is otherwise undermined. These debates about democracy, the public sphere and techno-social conditions of communication are increasingly being subsumed within a theory of community.

The levels argument explored in the previous chapter suggests that abstract communities have just as much intensity as face-to-face-based senses of community. Even the most abstract and disembodied levels have rituals and ceremonies capable of integrating 'strangers' – indeed ordinarily they transform strangers into natives. These rituals may be symbolic, such as voting, watching sport, etc., but they can also be very practical, such as annual tourism pilgrimages, downloading email or watching the same programme, such as the evening news, every day of the year.

However, what distinguishes broadcast forms of ritual from media ritual in general is *publicity*, an element of display and of spectatorship. Publicity and public spectacle are, of course, an aspect of *flânerie* which cannot be achieved on the Internet. Remember that *flânerie* involves a dialectic of seeing and being seen. Display, rather than an exchange of sign-values and texts, is at its core. Broadcast institutionalizes such practices of spectacle display and spectatorship, which may or may not be technologically extended, by creating *specular* spaces of association.

This function of broadcast is posited by some sociologists as contributing to the maintenance of social order itself, as Bob Mullan has pointed out. Quoting from Shils and Young's well-known discussion of the coronation of Elizabeth II (1953), Mullan shows the continuity between the embodied and extended rituals of coronation, both of which sustain social order and power:

> For Shils and Young: 'the central authority of an orderly society, whether it be secular or ecclesiastical, is acknowledged to be the avenue of communication with the realm of the sacred values. Within its society, popular constitutional

monarchy enjoys almost universal recognition in this capacity, and it is therefore enabled to heighten the moral and civic sensibility of the society and to permeate it with symbols of those values to which the sensitivity responds. Intermittent rituals bring the society or varying sectors of it repeatedly into contact with this vessel of the sacred values. The Coronation provided, at one time and for practically the entire society, such an intensive contact with the sacred that we believe we are justified in interpreting it as ... a great act of national communion.' (cited in Mullan, 1997: 4)

Television, and, we could add, the press and radio are able to extend this communion to so many more than is possible in mutual spectatorship:

> On 2 June 1953 much of Britain came to a standstill as millions watched the regal ceremony from Westminster Abbey. In the weeks before the event 500,000 sets were sold as 'Coronation fever' swept the land. Despite the fact that there were only two million or so sets in existence somehow 20 million managed actually to watch the occasion. At the time the Coronation became the 'biggest event in television history' and was broadcast in France, Holland, Germany. It is estimated that ultimately the world audience measured some 277 million. (5–6)

But, as we saw with Carey's discussion of ritual in the previous chapter, celebrative communion is only one variant of ritual display. Carey (1998) also discusses rituals of shame, degradation and excommunication. In media past and present there are a plethora of examples to demonstrate how the depiction of 'acts of cruelty can promote, however distastefully, states of social integration' (43). Drawing on Dayan and Katz (1992), he lists the McCarthy, Watergate and Iran–Contra hearings, and his own case study of the Robert Bork Senate hearing as a 'media event'. To these political rituals must be added the confessional forms of TV talk shows, and the tribal cruelty of excommunication that occurs on reality survivor shows.

Like the ceremonial appropriation of ritual, the ritual of shaming and exclusion also has historical antecedents that operated prior to electronic media. However, unlike the restitutive and deliberative 'ordeals' of modern democracy, the older regimes of ritual were punitive and violent. The most common occasion for display in pre-media societies was for torture, execution and 'punishment'. As John B. Thompson (1995) demonstrates (see also Chapter 2), in societies of the ancient world, the organization of power was based on the visibility of the few by the many. This could take the form of the visibility of nobility via the royal progression, the coronation or funeral, but also the careful management of public punishment in towns and cities throughout a nation.[27]

However, there are three major differences which Thompson outlines between the mediation of visibility by broadcasting technology, and the spectacle of old. Firstly, there is the range and scope, which we have already illustrated with the coronation example. Secondly, there is the 'field of vision', which enables individuals to see things that are 'far removed from the locales of their day-to-day lives' (130). Finally, there is the 'directionality': 'In the case of television … the direction of vision is essentially one-way. … The kind of publicness created by television is thus characterized by a fundamental contrast between producers and recipients in terms of their visibility and invisibility, their capacity to see and to be seen' (130).

Thompson's chronicle of broadcasting links mass media to a new form of publicness based on visibility. He suggests that such a publicness does not require speech in the Habermasian sense, but rather allows individuals to assume degrees of visibility and invisibility. Even if they are part of an *agora* in which they are invisible and have no speech, they can at least feel integrated with those who share their situation, and who are metaphorically represented in the serial world of the screen. In the latter sense, all audiences of mass media have visibility, in the very liveness of the medium. On the Internet, however, everyone has speech, but, in Akio's third register, no visibility at all.

Tanjev Schultz (2000) is one writer who wishes to reclaim the function of mass media as an agent of integration and as 'providers of a shared lifeworld'. He sees 'a future for this very function'.

> Criticism of mass media power and centralization does not necessarily deny their immense achievements. Jürgen Habermas, for instance, continued the cited tradition of worry about lack of interaction in mass media …. But he also viewed their integrational role as a benefit of modernism and as a necessary condition for a vivid public sphere in complex societies. (Schultz, 2000: 208)

For Habermas, however, visibility in itself is an insufficient basis for the public sphere if it does not include speech. At the same time, the power of speech is attenuated if little is known of the speaker, as Habermas also points out in *The Theory of Communicative Action* (1984). However, the ritual function of media is absent from Habermas's theory. Regardless of what kinds of texts circulate within mass media, they provide a place in which people construct meaning in their lives. Martin-Barbero (1997) suggests that we should look for 'the processes of re-enchantment in the continuing experience of ritual in communitarian celebration and in the other ways that the media bring people together' (108).

Not only does television provide such ritual opportunity, but it also has an enduring constancy. Ruth Rosen (1986) refers to a study by Agnes Nixon which found that most viewers (of soap opera in particular) regarded TV as the only 'constant' in their lives. The world outside may be in chaos and flux but TV provides an anchor. 'All potential viewers are members of a society that has been in constant transformation through geographic mobility and the loss of extended families' (Rosen, 1986: 46) find solace in tuning in to characters who, regardless of the convulsions in their life, are always with you Monday to Friday. Thus the ritual involves a constancy of participation, not just the consumption of texts.

As Baym (2000) suggests, 'Being a member of an audience community is not just about reading a text in a particular way; rather, it is about having a group of friends, a set of activities one does with those friends, and a world of relationships and feelings that grow from those friendships' (207). Such friendship may be with the characters as much as with other 'audience friends' who are enveloped by the same form of virtual participation.

This densely textured world of participation provides a standing reserve of sign-values, a reserve which is at its richest in the wider bandwidth of broadcast, which is, ultimately, television. Television is able to convey complexity but with a remarkable uniformity across a population.[28] Never have so may complex meanings – whether these be readerly or writerly in the Barthesian sense, hot or cool in the McLuhanist sense – been available to such large audiences.

Following our earlier discussion, the definition of the *flâneur*, whether embodied or electronic, as one who bathes in the crowd can be satisfied within both audience and network environments. However, it can be observed that audience–medium interactions offer much more to virtual *flâneurs* than do network–medium interactions.

As we saw with the rise of the *flâneur*, the excitement of *flânerie* was not a return to *Gemeinschaft* and community organized around interaction between members of a speech community. Rather, it was about interaction with a crowd of some form. Few physical forms of mutual association outside of broadcast assemblies offer this effect today save for the spectacle of the large sporting event.

The media offer the opportunity for people to come together to understand the central questions of life, from the meaning of art to the meaning of death, of sickness, of youth, of beauty, of happiness and of pain (see Martin-Barbero, 1997).

But audience communities which are organized around texts do not fully account for the kinds of social integration that are possible within broadcast. We have seen already how practices of media usage are common to network and broadcast dynamics alike. Broadcast, which may be said to 'influence consciousness', is also an environment for practices and rituals which are not simply semiotic.

Symbolic inequality in broadcast communities

The constancy that is provided by media genres to provide a common culture over time points to media as a mythological 'centre' of social life in modern societies, to use Couldry's (2003) phrase, but it is a constancy which is uneven in its broader social expressions. Whilst, as a general rule, the production of visibility is the foundation of audience communities, the genre and content of a media product can produce differing affects, divisions and identifications that stratify audience communities.

As argued, broadcast may be a form of reciprocity without interaction, where the many speak to the many, but they do so by way of agents of communication, media presenters, hosts, celebrities and 'personalities' who become intermediaries enabling such reciprocity to occur. This is expressed in a very weak level of reciprocity versus a strong degree of identification with those celebrities. These persons become 'media friends' (Meyrowitz, 1994). Such a world of virtual friendship is usually totally separate from the kind of friendships we might have in face-to-face or CMC forms. Only rarely do these worlds meet, as when we actually meet a celebrity, and there is a very dramatic collapsing of these worlds into each other. But the places that become famous in broadcast media can also become physically sacred, to the point where fans and audiences go on pilgrimage to them (see Couldry, 2003: Ch. 5). Alternatively, there are cases of those who already live in a location made famous by a television series wanting to redefine the name of their place to match the fictional name featured in the programme.[29]

In a demonstration of the power that such agents have in social integration, Nick Couldry (2003) provides a list that is well suited to understanding this symbolic inequality:

- people calling out as their presence 'on air' is acknowledged (the studio chat show host turns to them and asks them to clap, 'show what they feel')
- people either holding back, or rushing forward, at the sight of a celebrity
- people holding back before they enter a place connected with the media, so as to emphasise the boundary they cross by entering it
- performances of media people that acknowledge their own specialness before a crowd of non-media people
- performances by non-media people in certain types of formalised media context, such as a talk show. (52)

These kinds of events indicate the extreme symbolic differentials between celebrities and so-called 'ordinary' individuals. Here the relationship between a 'larger-than-life' Big Subject and the infinity of small subjects is easily expressed in the idea that a person could become a 'household name'. But equally, it might be expressed in the practice of 'name-dropping', of an 'ordinary' person claiming to know a celebrity in some way. By finding a way to share in the persona of persons in whom mass recognition is concentrated, an ordinary person can claim some of this recognition by way of the celebrity as agent.

For the celebrity, on the other hand, name-dropping is impossible. To name-drop would be to place oneself lower on the hierarchy of recognition.

Instead celebrities refer to each other as though they make up the entire population. The familiar statement of the media celebrity concerning parties that 'everyone will be there' or 'everyone is here' implies that the celebrity class posits itself as standing in for all classes.

Those who are locked outside this celebrity class must find ways of sharing in it at-a-distance in ways discussed above, or accept their exclusion by participation in ritual attention to the anomic condition. The contemporary talk show is an example of this, fixated with unhappy and unsatisfied guests who, staged or otherwise, confess their loneliness, or feelings of rejection at losing a partner (see below for an extended discussion).

Horton and Wohl (1956) argue, based on the American experience, that insofar as the production of anomie is, in part, internal to media, the anomic class is recognized by the mass media themselves, and 'from time-to-time specially designed offerings have been addressed to this minority' (223). They give the example from 1951 of a very popular radio programme in the USA – 'The Lonesome Gal'. As they describe the programme, it sounds very much like a forerunner of reality television, which is, 'in fact, nothing but the reciprocal of the spectator's own parasocial role' (224). What they mean by this is that instead of being a managed or self-contained drama, 'The Lonesome Gal' presented a character who could potentially be any person in the audience, anyone 'ordinary' from the street. 'Her entire performance consisted of an unbroken monologue, unembarrassed by plot, climax or denouement. The Lonesome Gal simply spoke in a throaty, unctuous voice whose suggestive sexiness belied the seeming modesty of her words' (224). But most importantly the Lonesome Gal appealed to her audience as being 'only one of millions of lonely girls, seeking love and companionship', and in doing so, empathized with the low visibility that 'lonely heart' audience members might feel on a daily basis, but which the programme was actually helping reproduce. This basic structure of interchangeability between an audience member's aspirations to be recognized, and a *non-celebrity* role-model who occupies precisely such a place of recognition, is manifest in all forms of reality genres, which are discussed further below.

But the dominant fields of recognition from which all forms of broadcast recognition spring are based on a division between high and low visibility of personalities. Fame is, by definition, concentrated among a few, and the gap between the famous and the ordinary can be measured in economic terms by the endorsements, appearance and advertising fees actually paid to celebrities. More recently, it is also possible to measure the fame of a celebrity by websites, such as Fametracker, devoted to auditing, summarizing and documenting their fan support, covering the volume of and loyalty to numerous sites devoted to celebrity deification. There are also celebrity exchanges such as Celebdaq (UK) and Hollywood SX (USA) where imaginary currency is used to buy and sell celebrity stock. And celebrities themselves are using the World Wide Web to extend the market that they establish on the screen. Upon the pretext of revealing an inner personality, screen stars parasitize their own image to sell themselves in another medium for economic gain. Sites like WorldLive.Com host on-line diaries of megastars such as Melanie Griffiths, who receives hit-money from each of the millions of fans who log on to her diary. But also they might be enticed into buying jewellery from her 'Goddess' range of products. Such marketing is more powerful than an advertising campaign featuring the celebrity, as it proposes to overcome the symbolic inequality between fan and star by the star divulging common everyday feelings, but also by emulating a non-celebrity's only avenue for achieving fame – the personal web-page.

Lonely at the top   Because of the fact that celebrities belong to a very restricted visibility class, it is frequently difficult for them to associate with persons of low visibility. They relate to each other, in the form of a kind of 'status closure' that is bound up in image-making; many of them report the diffi-culty of having 'real relationships'. There is even a genre which provides for this fact, films where a member of the media elite (or high-visibility elite) wants to experience being a 'real person': for example, *Notting Hill*, *Coming to America*, and *Love Actually*.[30] Two strong themes emerge in these narratives. Firstly, there is the idea that a famous person needs to mas-querade as an 'ordinary' (perhaps anonymous) person, or at least down-play their fame, in order to see whether people appreciate them for reasons other than just being 'famous'. Secondly the celebrity finds the affections of others superficial and suspicious, and seeks some way in which they can view the 'true' behaviour of others.

But the 'true' behaviour of ordinary people must be experienced candidly, without their being aware of celebrities, cameras and micro-phones – a realm which is itself constructed by the culture industry and formalized in 'candid camera' genres of television. Opposite the celebrity, the ordinary media consumer cannot have access to an intimate or empathic connection with their media hero either. To do so would be to transgress a boundary that cannot be crossed in any open sense. According to Meyrowitz (1994), however, there is a restricted sense in which this becomes realized – where a fan accumulates a very private obsession with a celebrity, to the point that they become 'media friends'. In a culture where electronic relationships can become as real as physical ones, some categories of fans may come to confuse the two in strange ways, and live their physical life in relation to the screen identification. Fans become so obsessed with being like them that they take on their appearance, sur-round themselves with their iconic memorabilia, or may even stalk the same person, or, in the most extreme case of not being able to reconcile the virtual with the physical, threaten to kill them, or actually succeed in their murder (see Meyrowitz, 1994: 63–4).

Of course, when fans act out such behaviour 'fanatically' (which is the root of the term applied to them), they become not friends but others, as 'strangers' whose strangeness is inversely measurable alongside the fame of the star. This affirms not only the privacy of the celebrity, but also the boundary division between media and ordinary.

This division traps the star as much as the fan, shrinking the field of legitimate relationships to celebrity circles as their personality is so overdetermined by the image. As such, the celebrity must not only act on the set, but also act out a role as an heroic individual among other heroes. They are the last *flâneurs*, bathing themselves in the crowd.

Moreover, media influence the behaviour of non-celebrities who might appear in it. Individuals modify their behaviour on TV in terms of narratives of expectation.[31] This is as binding on the celebrity as it might be on a viewer. Paradoxically, the over-exposure of a celebrity via the image can become suffocating to the point that they feel they do not get any attention outside of playing the glamorous roles that are expected of them. On occasions this may result in being attracted to scandal for its own sake, or, in some recent cases, petty shoplifting – Jennifer Capriati and Winona Ryder.

Ultimately, the position of the celebrity can be quite fatalistic, their immersion in their own image leaving little room for escape, an index of which is the high rate of suicide among celebrities. Such suicide may also affect the behaviour of fans, who may become deeply melancholic or even commit suicide themselves insofar as they are enveloped by the pathos of their object-fixation.[32] A study by Wasserman (1984) showed that reported celebrity deaths by suicide are more likely to be imitated than are non-celebrity suicides. Alternatively, obsessive object relations with celebrities may result in some order of resentment towards the inequality of the culture industry. Celebrities hire security guards for their protection, and some of those who want to walk the streets and be like everyone else might end up paying the price, as John Lennon did.[33]

Rituals of audience community – metonymous identification

Whilst the celebrity might swing between the heroic and the melancholic, audiences have recourse to several forms of narrative through which to 'live' their relationship to broadcast culture. After all, they do have choice over the programming and genres of media events in which they participate.

In the following discussion, focusing on television audiences, I will describe some of these genres in terms of the way they bring about variations of audience community, but also some ways in which these medium-communities become consummated through behaviour on the Internet. In nearly all of the formats and genres I examine, the main ritual being enacted by television audiences is to identify with what is happening in a television studio metonymously. That is, it is as though what happens in the studio is somehow able to substitute for a larger reality beyond the screen. A field of recognition is established by which the activity in the studio metonymically stands in for, is substituted for, a reality which cannot be otherwise represented, such as the audience itself.

Television news is an important starting point for understanding the temporal life of audience communities. Whether on free-to-air or cable, television news functions as a hub around which other kinds of programming are organized. The major evening bulletin provides a time-mark for the end of the working day and the beginning of a sense of private control of one's pleasure. This time-mark is itself heavily promoted by the networks in the expectation that news service loyalty will lead to channel loyalty for the rest of the evening. Freeway billboards and internal promotion aggressively market news programming for its comprehensiveness, its parochial expertise, or simply because the presenters appeal to the image of a surrogate family who will look after the viewer's needs.

The format of news itself provides comfort to audiences in the very regularity of its structure. On commercial media this format usually involves some variation of a proven sequence, beginning with stories that have photo opportunities, car crashes, local politicians, national then international sound bites, followed by a three-minute world news round-up, sport, weather and the mandatory human interest story in conclusion.

The predictability of such programming, and the fact that its affective pattern shows little variation, from violence-rage, to public figures offering a voice of reason, to the closure of a cat rescued from a tree, offers a constancy in viewers' lives that may compensate for the disorder of the working day. Thus, the performativity of the news is carried by both its formulaic sequencing and its serialization from day to day.

No other television genre quite has this constancy and availability, which is why news is invariably a flagship for channel loyalty. The point here is that this quality of *constancy* is evident independently of the actual content of news texts. That narrative details change each day reaffirms the durability of the genre itself. Thus the ritual of television news is not related to any kind of bardic 'textual' function of representation (see Fiske and Hartley, 1978), but to the fact of its performance and the authority of this performance.

The metaphysical grounding of such authority is a doctrine of representation as vision, in which language is a 'system of representation' rather than a form of activity (see Carey, 1989: 80). But also, news is invested with an authority by which its producers are bestowed with the power to metonymically manage, codify and organize the representation of the world for us. The world is substituted by a referential format that 'stands in' for some larger, unknown reality.

As a visual, ephemeral medium, television news elevates the doctrine of expressive realism to a high point that is expressed in the fetish of the 'photo opportunity'. The ephemeral dimension which distinguishes TV news from the press is its preoccupation with 'live' footage. The expectations of 'liveness', which are in conflict with the regulated timetabling of television news, is accommodated by the continuous obsession with updates of news, sport and weather. But what is concealed in the announcement of 'breaking news' is the way in which all broadcasted news is live, in a way that news based on information retrieval is not. Almost all broadcast news services all over the world have a mirror site on the World Wide Web, but none of these sites have a specular status, except insofar as they are consumed as a substitute for a centre of broadcast.

The fact that television news cultivates a synchronous relation with audiences makes possible the determination of *every* news announcement as an event. The event is the activity of the broadcast itself, not the representation of something anterior to the broadcast. This is something that news shares with all 'events' in the media.

However, an everyday appreciation of this is often obscured by the fact that the routine quality of media is 'programmed' ahead of time on the part of producers, and within a representational ontology. It is, by definition, without spontaneity and instantaneity. Insofar as audiences live their relation to programming within a process model, such an aesthetics of reception governs their experience of television events.

The live event But television programming becomes ordinary and routine, except in relation to the possibility that it may be interrupted. In such a case the extraordinary media event itself can become a genre which claims a double status of 'liveness', live in relation to a non-media reality, and live in relation to a surprising departure from programming. As Dayan and Katz (1992) theorize this media form:

> The most obvious difference between media events and other formulas of genres of broadcasting is that they are, by definition, not routine. In fact, they are *interruptions* of routine; they intervene in the normal flow of broadcasting in our lives. Like the holidays that contrast with daily everyday routines, television events propose exceptional things to think about, to witness, and to do. Regular broadcasting is suspended and preempted as we are guided by a series of special announcements and preludes that transform daily life into something special and, upon the conclusion of the event, are guided back again. In the most characteristic event, the interruption is *monopolistic*, in that all channels switch away from their regularly scheduled programming in order to turn to the great event, perhaps leaving a handful of independent stations outside the consensus. Broadcasting can hardly make a more dramatic announcement of the importance of what is about to happen. (403–4)

The authority of the broadcast medium is never more heightened than in the interruptive live event, yet it is also more involving of audiences who may feel a special connection precisely in the singularity and distinctiveness of the event. The event, which can be as banal as a breakdown in transmission followed by an announcement of the 'return of *normal* programming', provides a time-stamp that stands out from the 'pounding' of regular programming.

The talk show The genre of the television talk show with a studio audience introduces another source of 'liveness' which adds a layer of ritual that reinforces a depth of feeling in audience communities. Talk shows like those of Jerry Springer which involve the audience, or Trisha, stand at the intersection of two forms of association: namely a physical assembly which acts as the content of an electronic assembly.[34] Both forms of assembly are asymmetrical, with most participants having little opportunity to speak compared to hosts and select guests. The two audiences become overlaid in interesting ways. The embodied audience is live at the point of production and consumption, whereas the electronic audience is usually only live at the point of consumption. However, by way of identification with the embodied audience, the electronic audience is able to feel as though they are there, involved with the proceedings as much as the studio audience they are identifying with. As McLaughlin (1993) suggests in her review of the work of Carpignano:

> The presence of the public on television 'produces a short circuit in the dichotomy' between textual production and reception. The studio audience participates in both the viewing of the text and its scripting, while the home viewer 'monitors a space where a negotiation of textual meanings is in progress much in the same way as his [sic] personal negotiation with the screen'. 'The act of viewing a text becomes an act of viewing an act of viewing'. (45)

   Whilst the television audience clearly identifies with the studio audience and both audiences identify with the talk show host, the host typically lacks the spatial authority that is bestowed on film stars, news presenters, pop stars or any figure who occupies a form of stage. As McLaughin observes, the talk show host mingles with the audience, and has the role of an intermediary rather than an expert (45). The genre is one of reversing the power relations between stage and audience via a metaphoric displacement of the studio audience by the TV audience. The studio audience is 'literally' on centre stage. The show is constructed around the audience and defined by their involvement. The studio audience is supposed to exhibit forms of folk knowledge and home truths which are privileged over any expertise on the part of the host or guest commentators.

   Thus, the talk show is a prominent example of the way in which broadcast enables forms of *reciprocity without interaction* between a large number of persons who are nevertheless profoundly involved with the affective identifications between the audiences, guests, hosts and companions in the viewing or studio experience. These identifications are, I argue, largely *metonymic*.

   At the individual level, the field of recognition of the talk show provides for the possibility of intimacy with very large audiences as well as a sense that an ordinary person may become a 'representative' performer on behalf of those millions of viewers who will never get that chance: 'Talk shows make sense for many performers, as means of dealing with their normal "invisibility". They are, following Thompson, "struggles of visibility" – a matter of "being seen before the social gaze, before a representative sample of the social body"' (Couldry, 2003: 118). The appearance of people on talk shows individually overcomes the problem of 'ordinariness' that television audiences suffer whilst more generally reinforcing the division between ordinary and media culture. For the individual involved, it is 'less a comment on how trashy they are and more a comment on the exclusiveness of television and the limited access of ordinary people to media representation' (Gamson, cited in Couldry, 2003: 118).

But talk shows also provide for, indeed excel in, the spectacle of shaming – the modern-day confessional box. Couldry (2003) argues that 'whatever the artificiality and indeed cruelty of such shows and their attendant ethical problems, part of their significance *for performers* derives from the opportunity they represent, against the odds, to be seen before a public audience, to emerge from invisibility' (118). Given how much the modern individual is atomized and physically sequestered from others, Couldry argues that we become very accepting of 'action at a distance' with, or on behalf of, others. But, he suggests, 'the price of the expansion of the boundaries of private experience, if indeed that is what is occurring, is to submit that experience to the power dimensions of the mediation process' (116). These power dimensions are shaped by the theme of the talk show (shaming, celebratory, etc.), the field of recognition that the host encourages between the guests and the two audiences, the spatial authority that the guests are given on the stage, and numerous other factors which the guests have no control over. This careful stage-management in turn limits the 'reality effect' of the show, in relation to which guests are encouraged to break out of the boundaries of their 'visibility trap' (125). When the guests do this, their actions are not just 'real' but are part of the '*really* real'. The '*really* real', Couldry explains, 'is the moment when something "genuinely" uncontrolled happens in the highly controlled setting of the studio' (125). Thus, it is the display of emotion, in the form of tears or violence between confessional subjects of a talk show, that receives its impact precisely because it is the opposite of the controlled, linear, composed production values of nearly all television formats.

Paradoxically, the communication of 'emotion' in this way, as something that television is otherwise incapable of conveying in its scripted genres, also appeals to being able to 'represent' the emotions of viewers, whose participation is displaced and metaphoric. But their identification is potentially far more powerful than that which they may have with a celebrity. For a start it may be cathartic that, finally, an 'ordinary person' is able to make their feelings known on air. There may be a sense of justice for the viewer also. Now 'we can hear *our* side' of the story rather than the envy-sponsored preoccupations of the rich and famous. Then there is the amplification of the reality effect that results from the fact that, whilst the non-celebrity is expected to perform, they are perceived to have no special agenda to do with some image they are trying to cultivate, which is the skill of celebrity. Instead, a field of identification is established where we 'really get to know them for who they really are'.

Reality television The talk show is an important forerunner of reality TV, which institutionalizes a cluster of practices by which the symbolic inequalities between media and 'ordinary' culture can be redressed. But this again is only from the standpoint of guests, who feel they are able to act as representatives of their 'ordinary' colleagues, and for individual viewers, who might identify a guest as 'standing in' for them in some way. Nevertheless, reality TV provides for forms of reciprocity, again by metonymous identification, which operate without the need for direct interaction.

It ought to be pointed out that the enormous popularity of reality television formats since the mid-1990s coincides with the rise of the Internet as a medium which, in McLuhanist terms, has reworked the dominant medium of television. Simply put, reality TV is a genre in which the audience *appears* interchangeable with the producer. In a media land-scape where individuals might expect greater visibility by dint of the possibilities of self-publishing on the Internet, so too this 'struggle for visibility' demands greater audience participation in traditional broadcast media.

Of course, the appearance of interchangeability is all it is. It is not possible for the *whole* of the audience to be so exchanged, only a random selection of that audience. But if the majority of an audience identify with persons who are seen to be legitimate representatives, the exchange takes on a convincing, even exciting, quality. This is because, as James Carey (1989) suggests:

> In our time reality is scarce because of access: so few command the machin-ery for its determination. Some get to speak and some to listen, some to write and some to read, some to film and some to view ... there is not only class conflict in communication but status conflict as well. (87–8)

What is also illusory is the idea that a reality TV show such as Endemol Corporation's thirty-seven-country formula *Big Brother* is some-how 'raw' and 'unscripted', whereas its narrative is so one-dimensionally determined by the gaze of the camera, architecture and editing. Whichever version one turns to – Dutch, Australian, French – the same kinds of cloistered interactions are developed, along with the same processes of othering – shaming, heroic adulation, sympathy.

These three elements – camera gaze, architecture and editing – combine to produce a peculiar effect in television convention, the inauguration of surveillance as a mediated spectacle (Andrejevic, 2004: 2). A distinct field of recognition is established in such programming by which audiences, via metonymous identification, watch themselves being watched in a form which is a potential *mis-en-abyme*. It is possible for a spectator to become a participant and vice versa – such that being in front of a camera and being in front of a screen become interchangeable.

In his book *Reality TV: The Work of Being Watched* (2004), Andrejevic cites an example of being both participant and spectator at once. He introduces the case of a web-cam artist, Anna Voog, who once placed on-line video of herself watching 'Big Brother' on a lazy Saturday. In this instance, Andrejevic argues, Voog is

> caught between her television and her camera. ... On the one hand is the promise of interactivity – that access to the means of media production will be thrown open to the public at large, so that 'everyone can have their own TV show'. ... On the other hand is the reality represented by reality TV – that interactivity functions increasingly as a form of productive surveillance allowing for the commodification of the products generated by what I call the work of being watched. (2)

For Andrejevic, the surveillance culture possible in reality TV works 'neatly as an advertisement for the benefits of submission to comprehensive surveillance in an era in which such submission is increasingly productive' (2). For Andrejevic, reality TV is not a democratization of television because it only permits the gates of participation to be opened once its subjects, including the viewers, have submitted to the authority of surveillence. This authority stamps itself on the legitimacy of other tele-mediated practices such as the web-cam. Voog's web-page itself highlights the way in which 'viewers themselves were increasingly *being* watched in the age of interactive media that have ostensibly ushered in an era of the end of privacy' (2).

The personal web-page   Andrejevic argues that personal web-cams and web-pages double as a home-grown version of reality TV. However, the control that we have over such images is seen to be emancipatory, unlike the highly regulated images that are wrought by the televisual medium. Thus, he ultimately endorses a second media age view that '[t]he internet allows people like [Voog] to become content producers, rather than remaining merely media consumers' and offers them 'the ability to control the product of their creative labour' (5).

Thus, in the face of mass-mediated symbolic inequality, the personal web-page breaks up the monopoly of the culture industry. This in turn is said to explain the very fashionability of private web-pages and web-cams. Cheung (2000), another theorist of this trend, argues for its emancipatory status as it 'allows ordinary people to present their "selves" to the net public'. Such emancipation is achieved by three means. Firstly, there is the size of the 'audience' that it can reach, which, whilst not on the same scale as that experienced by media celebrities, goes some way towards redressing the imbalance between the poles of broadcast circulation. Secondly, the bandwidth and ability to convey complexity in image, music and text allows for richer forms of 'impression management' than are achieved in face-to-face interaction. Thirdly, it is emancipatory because it 'insulates' the author from any embarrassment, and avoids the possibility of rejection that is experienced in mutual presence (Cheung, 2000: 49).

But Cheung does not explore the nature of web-page audiences, or deconstruct the idea that their authors are 'ordinary' and 'amateur' (43). The fact that such characterization is assigned already indicates the necessarily 'reactive' nature of such a practice. Which is to say, personal home-pages are not a derivative of Internet communication, but are in fact yet another ritual of *audience communities*.

To go back to Anna Voog, who manages *in practice* to make of her own person a viewer and a producer, such a convergence, which Andrejevic makes into a vignette of old and new media convergence, can be argued to have already been attained within the dynamics of broadcast architecture itself. Voog merely has recourse to the technological means of displacing the aura of the image onto the apparatus or means of communication. It is the apparatus which becomes reified, as the image becomes a metonymic condensation of the audience. The television audience can see themselves in such images, without making this the reflexive subject of a further broadcast on web-cam.

Telecommunity

> Electronically mediated communication to some degree supplements existing forms of sociability but to another extent substitutes for them. New and unrecognizable modes of community are in the process of formation and it is difficult to discern exactly how these will contribute to or detract from postmodern politics. (Poster, 1990: 154)

The term 'telecommunity' can be found in the text of Alvin Toffler's *The Third Wave* (1980). Without endorsing the historicism of this text, we can say that Toffler's description of technologically extended community as a *social* form is one that is useful in a general way.

Long before the Internet, Toffler points out that technological extension is a general feature of late-capitalist societies, but that the distinctive form of community which it makes possible is by way of the 'selective substitution of communication for transportation' (382). For Toffler, the dispersal of populations across cities, and between home and work, creates unnecessary anomie. When communications begin to replace commuting, he argues, it can actually revitalize face-to-face relationships insofar as it enables work-from-home (a prophecy of modern telework) where family bonds and time for neighbourhood bonds are enhanced.[35] At the same time, people who are shy or invalid can pursue community on the basis of interests – again a prophecy of how the Internet is actually used.

Of course, Toffler, like the second media age theorists of more than a decade later, saw the limitations of television in its lack of interactivity.[36] In line with the second media age thesis, he held great optimism for the ability of extended interaction to help create community, but was opposed to the idea of the 'electronic cottage' simply replacing other levels of community such as the face-to-face.

Toffler's enthusiasm for community being made across a range of intersecting levels can be better appreciated if we incorporate the insights of the ritual views of communication. To do so is to immediately recognize the value of both broadcast and Internet community as levels of community which can be negotiated with face-to-face levels of integration.

As we have seen in this chapter, virtual or telemediated community is possible within both broadcast and network architectures of communication. The rituals involved in each kind of community differ, as do the fields of identification that they produce. What they have in common, however, is the character of enabling participation at a distance, in a movement of expansion and contraction. Increasingly, the basis of consuming media retreats to private media spaces, and from this privacy, individuals are able to reach out to more global forms of connection, where those older intermediate forms of community have all but disappeared.

This double movement of media ritual, the expansion of public forms of display and visibility, but only from the interface of privately controlled spaces, marks a general change in the nature of 'interaction'. Between these forces, to interact with others is to interact with media. Such media receive their power from the fact they have a constancy that endures beyond any particular individual communication event. This is expressed in everything from the techno-spiritualism which worships divine communion in cyberspace, to the everyday micro-rituals of media consumption explored in this book.

Whether electronic mediated communication extends or substitutes for intermediate forms of community, this book has also argued that it can strongly be identified as a driver of urban and global culture which, whilst uneven in the way it is connected to the transmission of local cultures, nevertheless establishes quite distinct forms of culture itself, which have their own rituals. Studying these rituals in the coming decades and how they are related to social integration will be a task central to the social sciences.

Notes

Parts of this chapter are derived from reviews and conference presentations or proceedings which I have presented or published. These are: a review of Barney, Darin, *Prometheus Wired: The Hope for Democracy in the Age of Network Technology*, *The Australian Journal of Political Science*, 2001, 36 (3): 618–619, and Holmes, D. (1998) 'Sociology without the

Social? Governmentality and Globalisation', in *Refashioning Sociology: Responses to a New World Order*. Brisbane: QUT Publications, pp. 167–173.

1   Where the social exists outside the nation-state, it does so in 'supra-national' bodies (WHO, UN, etc.).

2   Of course, all of these nominated contenders for community can be considered 'imagined communities' in the Andersonian sense (Anderson, 1983). However, insofar as they are made possible by mediated publicness, and it is only though this *kind of publicness* that individuals gain access to these imagined communities, they are also lived as the really real.

3   The discursive formation of community, as a kind of 'intermediate' level of social integration, would, within a levels argument, fit well within the secondary, agency-extended levels of integration outlined in Chapter 5.

4   Anderson in *Imagined Communities* (1983) conjectures that one of the reasons for the stability of the nation-state is the 'remarkable confidence of community in anonymity' (40).

5   Whether it is about tuning in to the same radio or television time slot, or adopting the newspaper as our 'morning prayer', as Hegel once suggested, or visiting the same bookmarks on our web-browser, the interface of which itself has a familiar and reassuring pixilated architecture, or whether we are at home at the cybercafé, all of these places are practised to the point of a uniformity which can be monumental in character. One can relate to the standardization of media architectures like a web-browser or a news performance in the same way as monuments might become references for a traveller.

6   For an analysis of the physical and architectural qualities of these spaces, see Holmes (2001).

7   An alternative to the user perspective in self/technology relations is provided by Steven Johnson in his idea of 'interface culture', which is measured by the degree that aesthetic values are a part of a technological environment. It is not simply a matter of computers and other hardware/software configurations being 'user-friendly'. Rather, the 'computer must also *represent itself* to the user, in a language the user understands' (Johnson, 1997: 14).

8   For example, educationalists are interested in whether classroom environments can keep up with 'cyberspace'. Moursund (1996) posed the question of the rate of change in cyberspace to a sample of fifty administrators, who thought that changes in the dynamics and modes of possibility in cyberspace were about eighteen times faster than in embodied space (4).

9   Computer companies certainly are interested in developing their own historical mythology and aura around their products. For example, every piece of software from Microsoft Corporation is accompanied by a certificate of authenticity which is printed on cloth that has the image of 'Augusta Ada Byron, collaborator of Charles Babbage in the nineteenth century'. Ada is also trademarked by the US Department of Defense. Ada is the name of its Proprietary Programming Language (see Plant, 1998: 14–22).

10   From Plato's *Republic*, to Saint-Simon, Thomas More and William Morris, this tradition has been a powerful one in the West.

11   The grand discourse of the *perils* of the Internet is that of the super-panopticon, which itself has a genesis myth – that of the Library of Babel – 'of the universe as a repository of information' (Whitaker, 2000: 48) The use of the Internet as an encyclopedic basis for surveillance presents ever greater risks to privacy the more information comes to mediate all categories of activity. The accumulation of information also makes possible an enlightenment in reverse.

12   Whilst he does not acknowledge the fact, Tofts is here replicating the point which the philosopher Jacques Derrida makes about difference in language as constituted in the last instance by *language-as-writing*, in which the mark or the gram within a signifier is the minimal basis of conceptuality. Thus his invention of *différance* as a replacement of the French *différence*, where the 'a' is silent when spoken and is noticeable only in its written form. See Derrida (1986).

13   'Interactivity has almost turned into a dull buzzword. The term is so inflated now that one begins to suspect that there is much less to it than some people want to make it appear' (Schultz, 2000: 205; see also Silverstone, 1999: 95).

14   The bourgeois *flâneur* is also a mythic hero of cultural studies (see Wark, 1999). The hero stays up late, lives in the inner city, can identify with those in poverty, but never has to suffer it. As we shall see soon, the electronic *flâneurs* of the Internet display the same characteristics: they aestheticize their sessions of surfing, they want to save everyone else, and substitute the Internet for the world, oblivious to the fact that 90% of the world does not use the Internet, and for most of them, it would not be of any assistance if they did. There is even a dandy form of such electronic *flâneurs*, who put up their own web-page with elaborate image, music and text.

15   It is worth noting Nikolai Gogol's introduction to his love story 'Nevsky Prospect', set in St Petersburg circa 1835. Here the street itself becomes the hero (cited in Berman, 1992: 195–6).

16   As with virtual *flânerie*, embodied *flânerie* was an interaction based on a contained information set. Whereas, on the Internet, what you know about others might be confined to images or texts, with the traditional *flâneur* it was the image.

17   Recall here numerous appraisals of the Internet as home to the avatar.

18   'The hero of modern life is he who lives in the public spaces of the city' (Tester, 1996: 5).

19   According to Paul Virilio (2001), the 'immediacy and instantaneity' of modern information present 'serious problems for contemporary society' (23–4).

20   This view is very common among Internet utopians. See, especially, Gauntlett (2000: 13), Rheingold (1994), Whittle (1996: 241ff).

21   Exemplary is Negroponte (1995), who argues that the real benefits of Internet sub-media are that they routinize asynchronous communication, making it possible to communicate with whom you want to when you want to, and you do not have to respond spontaneously to other human beings. The more one spends one's life on-line, the more this becomes a way of life (see 167ff).

22   A definition which echoes many of the terms found in microeconomic theory.

23   This binary they describe as Manichaean, 'duelling dualists who feed off each other, using the unequivocal assertions of the other side as foils for their arguments' (167).

24   One way of overcoming this ontological binary is to suggest, as Roger Silverstone (1999) does, that '*all communities are virtual communities*'. He explains: 'The symbolic expression and definition of community, both with or without electronic media, has been established as a *sine qua non* of our sociability. Communities are imagined and we participate in them both with and without the face-to-face, both with and without touch' (104).

25   But for an empirical study that does, see the Stanford 'Internet and Society Study' (Nie and Erdring, 2000), discussed in Chapter 4.

26   For Stoll, network interactions are 'superficial' (23).

27   The punitive function of visibility is not as common today, with the exception of 'public humiliation' TV shows. See below the discussion of talk shows.

28   'Television significantly constitutes a domain in which people ordinarily share experiences of the same complex social messages' (Livingstone, 1990: 1).

29   Here note the case of 'Northern Exposure' in Canada and 'Seachange' in Australia. 'Seachange' was an extremely popular public broadcast series about a small seaside community known as 'Pearl Bay', which was filmed at the town of Barwin Heads in southern Victoria. After the series had attracted a cult following in 1999, the residents of the town met at the town hall to discuss whether they should change the town's name to 'Pearl Bay'. An analysis was made of how beneficial such a change would be for tourism, which was already picking up. However, the series came to an end, and Barwin Heads has retained its name. Nevertheless, some of the sets in the film have been repurposed for media pilgrimages, such as 'Diver Dan's Shed', now a restaurant.

30   This genre is the obverse of a format in which the 'ordinary' person masquerades as part of an elite (*Sunset Boulevard, Dave* and *Desperately Seeking Susan*).

31 Langer (1997: 167) advances that TV and cinema have different personality systems, with the star system of cinema maintaining spectacle-at-a-distance and television providing an idiom of intimacy.

32 Or the reporting of celebrity assassinations. 'Following John Lennon's murder, a teenage girl in Florida and a 30-year-old man in Utah killed themselves. Their suicide notes spoke of depression over Lennon's death' (Meyrowitz, 1994: 64).

33 As Meyrowitz notes, for Mark David Chapman, the murderer of John Lennon, the killing unravelled a tragic negotiation with his alter ego. Chapman believed he was Lennon, and emulated him in every way. 'John Lennon Killed by Stranger' was the headline in December 1980, but Meyrowitz points out he was a close media 'friend'. 'In a sense, John Lennon was killed by the sinister side of the same force that makes millions still mourn him and other dead media heroes, a new sense of connection to selected strangers created by those modern media that simulate the sights and sounds of real-life interactions' (Meyrowitz, 1994: 63).

34 When exploring talk shows, it is necessary to acknowledge the variety of formats, including therapeutic, confessional, the studio debate between lay people or a panel of experts, and programmes that draw out conflict between friends and lovers. As Couldry (2003: 120) observes, however, whatever their content, all talk shows have an underlying ritual form, which is about legitimately entering a television space, to engage in a form of intimacy with a broad public which is not possible in any other forum.

35 Certainly this is true of some Net groups, as Willson (1997) narrates. It is worth noting that virtual community participants often feel the need to reinforce/complement their disembodied relations by simulating, at the level of ritual, more embodied or sensorial contacts. For example, participants on the WELL, a virtual community on the Internet, have regular face-to-face picnics and social gatherings. The participants develop a more complete understanding of each other at such gatherings (Rheingold, 1994: 21).

36 'The emerging info-sphere will make possible interactive electronic contact with others who share similar interest … such relationships can provide a far better antidote to loneliness than television as we know it today, in which the messages all flow one way and the passive receiver is powerless to interact with the flickering image on the screen' (383).

REFERENCES

Adilkno (The Foundation for the Advancement of Illegal Knowledge) (1998). 'Probing McLuhan', in *Media Archive*, New York: Autonomedia.

Adorno, T. (1954). 'Television and the Patterns of Mass Culture', *Quarterly of Film, Radio, and Television*, Vol. 8.

Adorno, T. and Horkheimer, M. (1993). 'The Culture Industry: Enlightenment as Mass Deception', in S. During (ed.), *The Cultural Studies Reader*, London: Routledge.

Agamben, G. (1993). *The Coming Community*, trans. M. Hardt, Minneapolis: University of Minnesota Press.

Alexander, J. (1986). 'The "Form" of Substance: The Senate Watergate Hearings as Ritual', in S.J. Ball-Rokeach and M.G. Cantor (eds), *Media, Audience and Social Structure*, London: Sage.

Alford, J. (1983). 'The Myth of False Consciousness', *Australian Left Review*, No. 84, Winter, pp. 6–9.

Althusser, L. (1971). *Lenin and Philosophy and Other Essays*, trans. Ben Brewster, New York: Monthly Review Press.

Anderson, B. (1983). *Imagined Communities*, London: Verso.

Anderson, B. and Tracey, K. (2001). 'Digital Living: The Impact (or Otherwise) of the Internet on Everyday Life', unpublished research, Suffolk, UK, Btax CT Research.

Andrejevic, M. (2004). *Reality TV: The Work of Being Watched*, Lanham, MD: Rowman and Littlefield.

Ang, I. (1991). *Desperately Seeking the Audience*, London: Routledge.

Ang, I. (1996). *Living Room Wars: Rethinking Media Audiences for a Postmodern World*, New York: Routledge.

Austin, J.L. (1962). *How to Do Things with Words*, Oxford: Oxford University Press.

Barney, D. (2000). *Prometheus Wired: The Hope of Democracy in the Age of Network Technology*, Sydney: University of New South Wales Press.

Barr, T. (2000). *Newsmedia.Com.Au: The Changing Face of Australia's Media and Communications*, St Leonards: Allen and Unwin.

Baudelaire, C. (1972). 'The Painter of Modern Life', in *Baudelaire: Selected Writings on Art and Literature*, trans. P.E. Charvet. Harmondsworth: Penguin, pp. 395–422.

Baudrillard, J. (1982). *Simulations*, New York: Semiotext(e).

Baudrillard, J. (1983). 'The Implosion of Meaning in the Media', in *In the Shadow of the Silent Majorities*, trans. P. Foss, J. Johnston and P. Patton, New York: Semiotext(e).

Baudrillard, J. (1988). 'Los Angeles Freeways', in *America*, trans. C. Turner, London: Verso.

Baym, N. (1995). 'The Emergence of Community in Computer-Mediated Communication', in S. Jones (ed.), *Cybersociety: Computer-Mediated Communication and Community*, London: Sage.

Baym, N. (1998). 'The Emergence of On-line Community', in S. Jones (ed.), *Cybersociety 2.0: Revisiting Computer-Mediated Communication and Community*, London: Sage.

Baym, N. (2000). *Tune In, Log On: Soaps, Fandom, and Online Community*, London: Sage.

Becker, B. and Wehner, J. (1998). 'Electronic Media and Civil Society', in *Proceedings of Cultural Attitudes Towards Technology and Communication*, 1st Conference, available at *http://www.it.murdoch.edu.au/~sudweeks/catac98/*

Bell, C. and Newby, H. (1976). 'Communion, Communalism, Class and Community Action: The Sources of New Urban Politics', in D. Herbert and R. Johnston (eds), *Social Areas in Cities*, Vol. 2. Chichester: Wiley.

Bell, C. and Newby, H. (1974). *The Sociology of Community: A Selection of Readings*, London: Cass & Co.

Bell, D. (1962). 'America as a Mass Society: A Critique', in *The End of Ideology: On the Exhaustion of Political Ideas in the Fifties*, New York: Free Press.

Bell, D. (1980). 'A Reply to Weizenbaum', in T. Forrester (ed.), *The Microelectronics Revolution: The Complete Guide to the New Technology and Its Impact on Society*, Oxford: Blackwell.

Bell, D. and Kennedy, B. (2000). *The Cybercultures Reader*, London: Routledge.

Belsey, C. (1982). *Critical Practice*, London: Methuen.

Benedetti, P. and Dehard, N. (eds) (1997). *Forward Through the Rearview Mirror: Reflections on and by Marshall McLuhan*, Ontario: Prentice-Hall Canada Inc.

Benedikt, M. (ed.) (1991). *Cyberspace: First Steps*, Cambridge, MA: MIT Press.

Beniger, J.R. (1987). 'Personalization of Mass Media and the Growth of Pseudo-communities', *Communication Research*, Vol. 14, No. 30, 352–71.

Benjamin, W. (1977). 'The Paris of the Second Empire in Baudelaire', in *Charles Baudelaire: A Lyric Poet in the Era of High Capital*, trans. Harry Zohn, London: New Left Books.

Bennett, T. (1982). 'Theories of the Media, Theories of Society', in M. Gurevitch, T. Bennett, J. Curran and J. Woollacott, *Culture, Society and the Media*, London: Methuen.

Berman, M. (1982). *All That Is Solid Melts into Air*, New York: Simon and Schuster.

Bianculli, D. (1996). *Dictionary of Teleliteracy: Television's 500 Biggest Hits, Misses, and Events*, New York: Continuum.

Bishop, B. (1999). *Global Marketing for the Digital Age*, Lincolnwood IL: NTC Business Books.

Black, D. (2001). 'Internet Radio: A Case Study in Medium Specificity', *Media, Culture & Society*, Vol. 23: 397–408.

Bolter, J.D. and Grusin, R. (1999). *Remediation: Understanding New Media*, Cambridge, MA and London: MIT Press.

Boorstin, Daniel J. ([1961] 1962). *The Image: Or, What Happened to the American Dream*, New York: Atheneum.

Bott, E. (1971). *Family and Social Network: Roles, Norms, and External Relationships in Ordinary Urban Families*, New York: Free Press.

Bourdieu, P. (1993). 'Public Opinion Does Not Exist', in *Sociology in Question*, trans. R. Nice, London: Sage.

Boyer, C.M. (1996). *Cybercities: Visual Perception in the Age of Electronic Communication*, New York: Princeton Architectural.

Brecht, B. (2003). 'The Radio as an Apparatus of Communication', in A. Everett and J.T. Caldwell (eds), *New Media: Theories and Practices of Digitextuality*, New York: Routledge.

Brosnan, M. (1998). *Technophobia: The Psychological Impact of Information Technology*, London: Routledge.

Buchstein, H. (1997). 'Bytes That Bite: The Internet and Deliberative Democracy', *Constellations*, Vol. 4, No. 2: 248–63.

Buzzard, K.S.F. (2003). 'Net Ratings: Defining a New Medium by the Old: Measuring Internet Audiences', in A. Everett and J.T. Caldwell (eds), *New Media: Theories and Practices of Digitextuality*, New York: Routledge.

Caldwell, J. (1995). 'Excessive Style: The Crisis of Network Television', in *Televisuality: Style, Crisis and Authority in American Television*, New Brunswick, NJ: Rutgers University Press.

Calhoun, C. (1986). 'Computer Technology, Large-scale Social Integration and the Local Community', in *Urban Affairs Quarterly*, Vol. 22, No. 2: 329–49.

Calhoun, C. (1992). 'The Infrastructure of Modernity: Indirect Social Relationships, Information Technology, and Social Integration', in H. Haferkampf and N. Smelser (eds), *Social Change and Modernity*, Berkeley: University of California Press.

Calhoun, C. (1998). 'Community Without Propinquity Revisited: Communications Technology and the Transformation of the Urban Public Sphere', *Sociological Inquiry*, Vol. 68, No. 3: 373–97.

Cantril, H., Gaudet, H. and Herzog, H. (1940). *The Invasion from Mars,* Princeton, NJ: Princeton University Press.

Carey, J.W. (1969). 'The Communications Revolution and the Professional Communicator', in P. Halmos (ed.), *The Sociology of Mass Media Communicators, Sociological Review Monograph 13*, University of Keele.

Carey, J. (1972). 'Harold Innis and Marshall McLuhan', in R. Rosenthal (ed.), *McLuhan: Pro & Con*, London: Pelican.

Carey, J. (1989). *Communication as Culture: Essays on Media and Society*, Boston: Unwin Hyman.

Carey, J. (1995). 'Time, Space and the Telegraph', in D. Crowley and P. Heyer (eds), *Communication in History*, White Plains, NY: Longman USA.

Carey, J. (1998). 'Political Ritual on Television: Episodes in the History of Shame, Degradation and Excommunication', in T. Liebes and J. Curran (eds), *Media, Ritual, Identity*, London: Routledge.

Casey, B. (2002). *Television Studies: The Key Concepts*, London: Routledge.

Cassidy, J. (2002). *Dot.Con: The Greatest Story Ever Sold*, New York: HarperCollins.

Castells, M. (1996). *The Information Age: Economy, Society and Culture*, Vol. 1: *The Rise of the Network Society*, Malden, MA and Oxford: Blackwell.

Castells, M. (2001). *The Internet Galaxy*, Oxford: Oxford University Press.

Chakhotin, S. (1939). *The Rape of the Masses: The Psychology of Political Propaganda*, trans. E.W. Dickes, New York: Alliance Book Corporation.

Chan-Olmsted, S. (2000). 'Marketing-Mass Media on the World Wide Web: The Building of Media Brands in an Integrated and Interactive World', in A.B. Albarran and D.H. Goff (eds), *Understanding the Web: Social, Political and Economic Dimensions of the Internet*, Ames: Iowa State University Press.

Chesher, C. (1997). 'An Ontology of Digital Domains', in D. Holmes (ed.), *Virtual Politics: Identity and Community in Cyberspace*, London: Sage.

Cheung, C. (2000). 'A Home on the Web: Presentations of Self on Personal Homepages', in D. Gauntlett (ed.), *Web.Studies: Rewiring Media Studies for the Digital Age*, Oxford: Oxford University Press.

Cooley, C.H. (1909). *Social Organization: Study of the Larger Mind*, New York: C. Scribner's Sons.

Corner, J. (1997). 'Media Studies and the "Knowledge Problem"', in T. O'Sullivan and Y. Jewkes (eds), *The Media Studies Reader*, London: Arnold.

Corner, J. and Harvey, S. (1996). *Television Times: A Reader*, London: Arnold.

Couldry, N. (2003). *Media Rituals: A Critical Approach*, London: Routledge.

Crang, M., Crang, P. and May, J. (1998). *Virtual Geographies: Bodies, Space and Relations*, London: Routledge.

Crowley, D. and Mitchell, D. (1995). 'Communication in a Post-Mass Media World', in D. Crowley and D. Mitchell (eds), *Communication Theory Today*, Cambridge: Polity.

Davis, E. (1998). *Techgnosis: Myth, Magic, Mysticism in the Age of Information*, New York: Harmony.

Dayan, D. and Katz, E. (1992). *Media Events: The Live Broadcasting of History*, Cambridge, MA: Harvard University Press.

Debord, G. (1977). The *Society of the Spectacle*, trans. F. Perlman and J. Supak, Detroit: Black and Red Publications.

de Certeau, M. (1988). *The Practice of Everyday Life*, Los Angeles: University of California Press.

Dempsey, K. (1998). 'Community, Experiences and Explanations', in A. Kellehear (ed.), *Social Self, Global Culture*, Oxford: Oxford University Press.

Derrida, J. (1976). *Of Grammatology*, trans. G.C. Spivak, Baltimore: Johns Hopkins University Press.

Derrida, J. (1978). *Writing and Difference*, trans. A. Bass, London: Routledge & Kegan Paul.

Derrida, J. (1981). *Positions*, trans. A. Bass, London: Athlone Press.

Derrida, J. (1986). *Margins of Philosophy*, trans. A. Bass, Brighton: Harvester Press.

Derrida, J. (1988). *Limited Inc.* Evanston, IL: Northwestern University Press.

Dery, M. (ed.) (1994). *Flamewars: The Discourse of Cyberculture*. London: Duke University Press.

Dery, M. (1995). 'The Medium's Messenger', *21.C. Previews of a Changing World*, Vol. 2: 8–12.

Dery, M. (1996). *Escape Velocity: Cyberculture at the End of the Twentieth Century*, New York: Grove Press.

Di Maggio, P., Hargittai, E., Neuman, W.R. and Robinson, J. (2001). 'The Internet's Effects on Society', in *Annual Reviews of Sociology*, Vol. 27, 307–336.

Dizard, W. (2000). *Old Media New Media: Mass Communications in the Information Age* (3rd edn), New York: Addison-Wesley Longman.

Dominick, J.R. (2001). *The Dynamics of Mass Communication: Media in the Digital Age* (7th edn), Boston: McGraw-Hill.

Donath, J. (1999). 'Identity and Deception in the Virtual Community', in M. Smith and P. Kollock (eds), *Communities in Cyberspace*, London: Routledge.

Downes, T. and Fatouros, C. (1995). *Learning in an Electronic World: Computers in the Classroom*, Newtown: Primary English Teaching Association.

Droege, P. (ed.) (1997). *Intelligent Environments: Spatial Aspect of the Information Revolution*. Amsterdam, New York: Elsevier.

Du Gay, P., Hall, S., James, L., Mackay, H. and Negus, K. (1997). *Doing Cultural Studies: The Story of the Sony Walkman*, London: Sage in association with the Open University.

Durkheim, E. (1982). *The Rules of Sociological Method*, ed. S. Lukes; trans. W.D. Halls, New York: Free Press.

Durkheim, E. (1984). *The Division of Labour in Society*, Basingstoke: Macmillan.

Eade, J. (ed.) (1997). *Living the Global City: Globalization as a Local Process*, London: Routledge.

Eagleton, T. (1991). *Ideology*, London: Verso.

Edge, D. (1988). *The Social Shaping of Technology*, Edinburgh: PICT Working Papers No. 1.

Ellis, J. (1982). *Visible Fictions: Cinema, Television, Video*, London: Routledge and Kegan Paul.

Elsaesser, T. and Hoffmann, K. (eds) (1998). *Cinema Futures: Cain, Abel or Cable? The Screen Arts in the Digital Age*, Amsterdam: Amsterdam University Press.

Escobar, A. (1994). 'Welcome to Cyberia', *Current Anthropology*, Vol. 5, No. 3, 211–31.

Ettema, J.S. and Whitney, D.C. (eds) (1994). *Audiencemaking: How the Media Create the Audience*, London: Sage.

Evans, J. and Hall, S. (eds) (1999). *Visual Culture: The Reader*, London: Sage.

Feather, J. (2000). *The Information Society: A Study of Continuity and Change* (3rd edn), London: Library Association Publications.

Featherstone, M. (1998). 'The Flâneur, the City and Virtual Public Life', *Urban Studies*, Vol. 35, Nos. 5–6: 910–18.

Featherstone, M. and Burrows, R. (eds) (1995). *Cyberspace, Cyberbodies, Cyberpunk: Cultures of Technological Embodiment*, London: Sage.

Feenberg, A. (1991). *Critical Theory of Technology*, Oxford: Oxford University Press.

Fidler, R. (1997). *Mediamorphosis: Understanding New Media*, Thousand Oaks, CA: Pine Forge Press.

Finnegan, R. (2002). *Communicating: The Multiple Modes of Human Communication*, London: Routledge.

Fiske, J. (1982). *Introduction to Communication Studies*, London: Methuen.

Fiske, J. (1987). *Television Culture*, London: Methuen.

Fiske, J. and Hartley, J. (1978). 'Bardic Television', in *Reading Television*, London: Methuen.

Flew, T. (2002). *New Media: An Introduction*, Melbourne: Oxford University Press.

Flitterman-Lewis, S. (1992). 'Psychoanalysis, Film and Television' in R.C. Allen (ed.), *Channels of Discourse, Reassembled: Television and Contemporary Criticism*, Chapel Hill, NC: University of North Carolina Press.

Foster, D. (1997). 'Community and Identity in the Electronic Village', in D. Porter (ed.), *Internet Culture*, London: Routledge.

Fraser, N. (1990). 'Rethinking the Public Sphere: A Contribution to the Critique of Actually Existing Democracy', *Social Text*, Vol. 25/26: 56–80.

Friedberg, A. (1993). *Window Shopping: Cinema and the Postmodern*, Berkeley: University of California Press.

Fukuyama, F. (1995). *Trust: The Social Virtues and the Creation of Prosperity*, New York: Free Press.

Fukuyama, F. (1999). *The Great Disruption: Human Nature and the Reconstitution of Social Order*, New York: Free Press.

Garofalo, R. (1991). 'Understanding Mega-events: If We Are the World, Then How Do We Change It?', in C. Penley and A. Ross (eds), *Technoculture*, Minneapolis: University of Minnesota Press.

Gates, B. (1996). *The Road Ahead*, London: Penguin.

Gauntlett, D. (ed.) (2000). *Web.Studies: Rewiring Media Studies for the Digital Age*, Oxford: Oxford University Press.

Gauntlett, D. and Hill, A. (1999). *TV Living: Television Culture and Everyday Life*, London: Routledge.

Geraghty, C. and Lusted, D. (1998). *The Television Studies Book*, London: Arnold.

Gerbner, G. (1956). 'Toward a General Model of Communication', *Audio Visual Communication Review*, Vol. IV, No. 3: 171–99.

Gergen, K.J. (1991). *The Saturated Self: Dilemmas of Identity in Contemporary Life*, New York: Basic Books.

Giddens, A. (1986). *Sociology: A Brief but Critical Introduction*, London: Macmillan.

Giddens, A. (1987). *A Contemporary Critique of Historical Materialism*, Berkeley: University of California Press.

Giddens, A. (1990). *The Consequences of Modernity*, Cambridge: Polity.

Gilder, G. (1993). 'The Death of Telephony: Why the Telephone and TV Will Not Be the Stars of a Communications Revolution', *The Economist*, 11–17 September 1993.

Gilder, G. (1994). *Life After Television*, New York: Norton.

Gitlin, T. (1998). 'Public Sphere or Public Sphericules?', in T. Liebes and J. Curran (eds), *Media, Ritual, Identity*, London: Routledge.

Gitlin, T. (2002). *Media Unlimited: How the Torrent of Images and Sounds Overwhelms Our Lives*, New York: Henry Holt and Co.

Goffman, E. (1973). *The Presentation of Self in Everyday Life*, Woodstock, NY: Overlook Press.

Goodheart, E. (2000). 'Marshall McLuhan Revisited', *Partisan Review*, Vol. 67: 90–100.

Gouldner, A. (1976). *The Dialectic of Ideology and Technology*, London: Macmillan.

Graber, D.A. (2001). *Processing Politics: Learning from Television in the Internet Age*, Chicago: University of Chicago Press.

Graham, S. (2000). *Splintering Urbanism: Networked Infrastructures, Technological Mobilities and the Urban Condition*, London: Routledge.

Graham, S. and Aurigi, A. (1998). 'The "Crisis" in the Urban Public Realm', in B. Loader (ed.), *Cyberspace Divide: Equality, Agency and Policy in the Information Society*, London: Routledge.

Graham, S. and Marvin, S. (1996). *Telecommunication and the City: Electronic Spaces, Urban Places*, London: Routledge.

Grajczyk, A. and Zollner, O. (1996). 'How Older People Watch Television: Telemetric Data on the TV Use in Germany in 1996', *Gerontology*, Vol. 44, No. 3: 176–81.

Green, B. and Bigum, C. (1993). 'Aliens in the Classroom', *Australian Journal of Education*, Vol. 37, No. 2: 119–41.

Greenfield, C. (1999). 'Home Alone? Mobile Privatization and the Transformation of Work', in J. Lee, B. Probert and R. Watts (eds), *Work in the New Economy: Policies, Programs, Populations*, Melbourne: Centre for Applied Social Research, RMIT University.

Guattari, F. (1986). 'The Postmodern Dead End', *Flash Art*, No. 128: 40–1.

Guest, A.M. and Wierzbiki, S.K. (1999). 'Social Ties at the Neighbourhood Level: Two Decades of GSS Evidence', *Urban Affairs Review*, Vol. 35, No. 1: 92–111.

Gumbrecht, H.U. and Marrinan, M. (eds) (2003). *Mapping Benjamin: The Work of Art in the Digital Age*, Stanford, CA: Stanford University Press.

Habermas, J. (1974). 'The Public Sphere: An Encyclopedia Article', in E. Bronner and D. Kellner (eds), *Critical Theory and Society: A Reader*, London: Routledge.

Habermas, J. (1984). *The Theory of Communicative Action* (Vol. 1), London: Heinemann.

Habermas, J. ([1962] 1989). *The Structural Transformation of the Public Sphere: An Inquiry into Bourgeois Society*, trans. T. Burger and F. Lawrence, Cambridge: Polity.

Habermas, J. (1992). 'Further Reflections on the Public Sphere', in C. Calhoun (ed.), *Habermas and the Public Sphere*, Cambridge, MA: MIT Press.

Hall, P. and Brotchi, J. (eds) (1991). *Cities of the 21st Century: New Technologies and Spatial Systems*, Melbourne: Longman Cheshire.

Hall S. (1977). 'Culture, the Media and the "Ideological Effect"', in J. Curram, M. Gurevitch and J. Woollacott (eds), *Mass Communication and Society*, London: Arnold.

Hall, S. (1980). 'Encoding/Decoding', in S. Hall, D. Hobson, A. Lowe and P. Wills (eds), *Culture, Media, Language: Working Papers in Cultural Studies, 1972–79*, London: Hutchinson.

Hall, S. (1982). 'The Rediscovery of "Ideology": Return of the Repressed in Media Studies', in M. Gurevitch, T. Bennett, J. Curram and J. Woollacott (eds), *Culture, Society and the Media*, London: Methuen.

Hall, S., Cruz, J. and Lewis, J. (1994). 'Reflections upon the Encoding/Decoding Model: An Interview with Stuart Hall', in J. Cruz and J. Lewis (eds), *Viewing, Reading, Listening: Audiences and Cultural Reception*, Boulder, Co: Westview.

Hanks, W.F. (1996). *Language and Communicative Practices*, Boulder, Co: Westview.

Haraway, D.J. (1991). 'A Manifesto for Cyborgs', in *Simians, Cyborgs and Women: The Reinvention of Nature*, New York: Routledge.

Hartley, J. (1992a). *Tele-ology: Studies in Television*, London: Routledge.

Hartley, J. (1992b). *The Politics of Pictures: The Creation of the Public in the Age of Popular Media*, New York: Routledge.

Hartley, J. (1998). 'This Way Lies Habermadness', *Media International Australia*, No. 88: 125–35.

Harvey, D. (1989). *The Condition of Postmodernity*, Oxford: Blackwell.

Hawisher, G.E. and Selfe, C.L. (2000). 'Introduction: Testing the Claims', in *Global Literacies and the World Wide Web*, London: Routledge.

Healy, D. (1997). 'Cyberspace and Place', in D. Porter (ed.), *Internet Culture*, London: Routledge.

Heidegger, M. (1997). 'The Question Concerning Technology', in *The Question Concerning Technology and Other Essays*, trans. W. Lovitt, New York: Harper Torchbooks.

Heller, S. and Anderson, G. (1994). *Typeplay*, Dusseldorf: Nippan Shuppan Hanbai.

Hepworth, M. and Ducatel, K. (1992). *Transport in the Information Age: Wheels and Wires*, London: Belhaven Press.

Herbert, T.E. and Proctor, W.S. (1932). *Telephony*, Vol. 1. 2nd edn, London: New Era.

Hills, M. (2001). 'Virtually Out There: On-line Fandom', in S. Munt (ed.), *Technospaces: Inside the New Media*, London: Continuum.

Hirst, P. (1976). 'Althusser and the Theory of Ideology', *Capital and Class*, Vol. 5, No. 4: 385–412.

Hirst, P. and Thompson, G. (1996). *Globalization in Question: The International Economy and the Possibilities of Governance*, Cambridge: Polity.

Holloran, J. (1970). *The Effects of Television*, London: Pantheon Books.

Holmes, D. (1989). 'Deconstruction: A Politics without a Subject, That Is', *Arena*, No. 88: 73–116.

Holmes, D. (ed.) (1997). *Virtual Politics: Identity and Community in Cyberspace*, London: Sage.

Holmes, D. (ed.) (2001). *Virtual Globalization: Virtual Spaces/Tourist Spaces*, London: Routledge.

Holmes, D. (2004). 'The Electronic Superhighway: Melbourne's CityLink Project', in S. Graham (ed.), *The Cybercities Reader*, London: Routledge.

Holmes, D. and Russell, G. (1999). 'Adolescent CIT Use: Paradigm Shifts for Educational and Cultural Practices?', *British Journal of Sociology of Education*, Vol. 20, No. 1: 73–82.

Horrocks, C. (2001). *Marshall McLuhan and Virtuality*, Lanham, MD: Totem Books.

Horton, D. and Wohl, R. (1956). 'Mass Communication and Para-social Interaction: Observation on Intimacy at a Distance', *Psychiatry*, Vol. 19: 215–29.

Howard, P.E.N., Rainie, L. and Jones, S. (2001). 'Days and Nights in the Internet: The Impact of Diffusing Technology', *American Behavioral Scientist*, Vol. 45: 383–404.

Huyssen, A. (1995). 'In the Shadow of McLuhan: Baudrillard's Theory of Simulation', in *Twilight Memories*, New York: Routledge.

Inglis, F. (1993). *Media Theory: An Introduction*, Oxford: Blackwell.

Innis, H. (1964). *The Bias of Communication*, Toronto: University of Toronto Press.

Innis, H. (1972). *Empire and Communications*, Toronto: University of Toronto Press.

James, P. and Carkeek, F. (1997). 'This Abstract Body', in D. Holmes (ed.), *Virtual Politics: Identity and Community in Cyberspace*, London: Sage.

Jameson, F. (1991). *Postmodernism, or The Cultural Logic of Late Capitalism*, London: Verso.

Johnson, S. (1997). *Interface Culture: How Technology Transforms the Way We Create and Communicate*, New York: HarperCollins.

Jones, P. (2000). 'McLuhanist Societal Projections and Social Theory: Some Reflections', *Media International Australia*, No. 94: 39–56.

Jones, S. (ed.) (1995). *Cybersociety: Computer-Mediated Communication and Community*, London: Sage.

Jordan, T. (1999). *Cyberpower: The Culture and Politics of Cyberspace and the Internet*, London: Routledge.

Jowett, G. (1981). 'Extended Images', in R. Williams (ed.), *Contact: Human Communication and Its History*, London: Thames and Hudson.

Kaplan, N. (2000). 'Literacy Beyond Books', in A. Herman and T. Swiss (eds), *The World Wide Web and Contemporary Cultural Theory*, London: Routledge.

Katz, E. Blumler, J. and Gurevitch, M. (1974). 'Utilization of Mass Communication by the Individual', in J. Blumler and E. Katz (eds), *The Uses of Mass Communications: Current Perspectives on Gratifications Research*, Beverly Hills and London: Sage.

Katz, J.E. and Aakhus, M.A. (eds) (2001). *Perpetual Contact: Mobile Communication, Private Talk, Public Performance*, Cambridge: Cambridge University Press.

Kearney, R. (1984). *Dialogues with Contemporary Continental Thinkers: The Phenomenological Heritage*, Manchester: Manchester University Press.

Kirby, V. (1997). *Telling Flesh*, London: Routledge.

Kling, R. et al. (2000). 'Anonymous Communication Policies for the Internet: Results and Recommendations of the AAAS Conference', in R.M. Baird, R. Ramsower and S.E. Rosenbaum (eds), *Cyberethics: Social and Moral Issues in the Computer Age*, New York: Prometheus Books.

Knorr-Cetina, K. (1997). 'Sociality with Objects: Social Relations in Postsocial Knowledge Societies', *Theory, Culture, Society*, Vol. 14, No. 4: 1–30.

Kornhauser, W. (1960). *The Politics of the Masses*, London: Routledge and Kegan Paul.

Krause, L. and Petro, P. (eds) (2003). *Global Cities: Cinema, Architecture, and Urbanism in a Digital Age*, New Brunswick, NJ: Rutgers University Press.

Kress, G.R. (2003). *Literacy in the New Media Age*, London and New York: Routledge.

Kroker, A. (1995). *Digital Humanism: The Processed World of Marshall McLuhan*, Montreal: CTheory.

Kroker, A. (2001). *Technology and the Canadian Mind: Innis/McLuhan/Grant*, Montreal: CTheory.

Kroker, A. and Kroker, M. (1996). *Hacking the Future: Stories for the Flesh-Eating 90s*, Montreal: CTheory.

Kroker, A. and Weinstein, M. (1994). *Data Trash: The Theory of the Virtual Class*, New York: St Martin's Press.

Kuo, D. (2001). *Dot.bomb: My Days and Nights at an Internet Goliath*, London: Little, Brown.

Lacan, J. (1985). 'The Agency of the Letter in the Unconscious: Or Reason since Freud', in *Ecrits: A Selection*, trans. A. Sheridan, London: Tavistock.

Lakoff, G. (1995). 'Body, Brain and Communication', an interview with I.A. Boal, in J. Brook and I. Boal (eds), *Resisting the Virtual Life: The Culture and Politics of Information*, San Francisco: City Lights.

Langer, J. (1997). 'Television's Personality System', in T. O'Sullivan and Y. Jewkes (eds), *The Media Studies Reader*, London: Arnold.

Lapham, L. (1994). 'Introduction to the MIT Press Edition: The Eternal Now', in M. McLuhan, *Understanding Media: The Extensions of Man*, Cambridge, MA: MIT Press.

Larrain, J. (1983). *Marxism and Ideology*, London: Macmillan.

Lasn, K. (2000). *Culture Jam*, New York: Quill.

Lasswell, H. (1948). 'The Structure and Function of Communication in Society', in L. Bryson (ed.), *The Communication of Ideas*, New York: Harper.

Lave, J. and Wenger, E. (1991). *Situated Learning: Legitimate Peripheral Participation*, New York: Cambridge University Press.

Lax, S. (2000). 'The Internet and Democracy', in D. Gauntlett (ed.), *Web.Studies: Rewiring Media Studies for the Digital Age*, Oxford: Oxford University Press.

Lazarsfeld, P.F. and Kendall, P.L. (1949). 'The Communications Behavior of the Average American', in W. Schramm (ed.), *Mass Communications*, Urbana: University of Illinois Press.

Lealand, G. (1999). Paper presented to the Australian Key Centre for Media and Cultural Policy, 'Regulation – what Regulation? Cultural Diversity and Local Content in New Zealand Television in the 1990s', September.

Leavis, F.R. (1930). *Mass Civilisation and Minority Culture*, Cambridge: Minority Press.

Lee, H. and Liebenau, J. (2000). 'Time and the Internet at the Turn of the Millennium', *Time & Society*, Vol. 9, No. 1: 43–56.

Lefebvre, H. (1991). *The Production of Space*, trans. D. Nickolson-Smith, Oxford: Blackwell.

Le Grice, M. (2001). *Experimental Cinema in the Digital Age*, London: British Film Institute.

Lemaire, A. (1970). *Jacques Lacan*, London: Routledge and Kegan Paul.

Levinson, P. (1999). *Digital McLuhan: A Guide to the Information Millennium*, New York: Fordham University/Routledge.

Lévy, P. (2001). *Cyberculture*, trans. Robert Brononno, Minnesota: University of Minnesota Press.

Liebes, T. and Curran, J. (1998). *Media, Ritual, Identity*, London: Routledge.

Lievrouw, L. and Livingstone, S. (2002). *Handbook of New Media: Social Shaping and Consequences of ICTs*, London: Sage.

Lipset, S. (1963). *Political Man*, London: Heinemann.

Livingstone, S. (1990). *Making Sense of Television*, London: Routledge.

Lowery, S. and De Fleur, M. (1983). *Milestones in Mass Communication Research*, New York: Longman.

Lukács, G. (1971). *History and Class Consciousness: Studies in Marxist Dialectics*, trans. R. Livingstone, London: Merlin Press.

Lukes, S. (1973). *Émile Durkheim: His Life and Work: A Historical and Critical Study*, Harmondsworth: Penguin.

Lull, J. (1995). *Media, Communication, Culture: A Global Approach*, Cambridge: Polity.

Lunenfeld, P. (1999). *The Digital Dialectic: New Essays on New Media*, Cambridge, MA: MIT Press.

Lupton, D. (1999). 'Monsters in Metal Cocoons: Road Rage and Cyborg Bodies', *Body and Society*, Vol. 5, No. 1: 57–92.

Lyotard, J.-F. (1984). *The Postmodern Condition*, trans. G. Bennington and B. Massumi, Manchester: Manchester University Press.

McCarthy, A. (2001). *Ambient Television: Visual Culture and Public Space*, Durham, NC: Duke University Press.

McChesney, R. (2000). 'So Much for the Magic of Technology and the Free Market: The World Wide Web and the Corporate Media System', in A. Herman and T. Swiss (eds), *The World Wide Web and Contemporary Cultural Theory*, London: Routledge.

McLaughlin, L. (1993). 'Chastity Criminals in the Age of Electronic Reproduction: Re-viewing Talk Television and the Public Sphere', *Journal of Communication Inquiry*, Vol. 17, No. 1: 41–55.

McLuhan, E. and Zingrone, F. (eds) (1997). *The Essential McLuhan*, London: Routledge.

McLuhan, M. (1964). 'Introduction' to H. Innis, *The Bias of Communication*, Toronto: University of Toronto Press, reprinted in E. McLuhan and F. Zingrone (eds), *The Essential McLuhan*, London: Routledge, 1997.

McLuhan, M. (1967). *The Gutenberg Galaxy: The Making of Typographic Man*, London: Routledge and Kegan Paul.

McLuhan, M. (1994). *Understanding Media: The Extensions of Man*, Cambridge, MA: MIT Press.

McLuhan, M. and Fiore, Q. (1967). *The Medium is the Massage: An Inventory of Effects*, London: Penguin.

McLuhan, M. and Fiore, Q. (produced by Jerome Agel) (2001). *War and Peace in the Global Village*, Corte Madera: Ginko Press.

McNeill, A. (1996). *Total Television*, London: Penguin.

McNeill, L. (1997). *Travel in the Digital Age*, London: Bowerdean Publishing.

McQuail, D. (1983). *Mass Communication Theory: A Reader*, London: Sage.

McQuire, S. (1995). 'The Go-for Broke Game of History', *Arena Journal* (New Series), No. 4: 201–27.

Marc, D. (2000). 'What Was Broadcasting?', in H. Newcomb (ed.), *Television: A Critical View*, New York: Oxford University Press.

Martin-Barbero, J. (1997). 'Mass Media as a Site of Resacralization of Contemporary Cultures', in S. Hoover and K. Lindby (eds), *Rethinking Media, Religion and Culture*, London: Sage.

Marx, K. (1973). *Grundrisse*, London: Penguin.

Marx, K. (1977). *Economic and Philosophical Manuscripts of 1844*, Moscow: Progress Publishers.

Marx, K. (1976). *Capital*, Vol. I, London: Penguin.

Marx, K. and Engels, F. (1970). *The German Ideology*, New York: International Publishers.

Mattelart, A. (2000). *Networking the World, 1794–2000*, trans. L. Carey-Libbrecht and J.A. Cohen, Minneapolis and London: University of Minnesota Press.

Mattelart, A. and Mattelart, M. (1992). *Rethinking Media Theory: Signposts and New Directions*, Minneapolis: University of Minnesota Press.

Mellencamp, P. (ed.) (1991). 'TV Time and Catastrophe, or Beyond the Pleasure Principle of Television', in P. Mellencamp (ed.), *Logics of Television: Essays in Cultural Criticism*, Bloomington: Indiana University Press.

Meyrowitz, J. (1985). *No Sense of Place: The Impact of Electronic Media on Social Behavior*, New York: Oxford University Press.

Meyrowitz, J. (1990). 'Television: The Shared Arena', *The World and I*, Vol. 5, No. 7: 464–81.

Meyrowitz, J. (1994). 'The Life and Death of Media Friends: New Genres of Intimacy and Mourning', in R. Cathcart and S. Druckers (eds), *American Heroes a Media Age*, Cresskill, NJ: Hampton Press.

Meyrowitz, J. (1995). 'Medium Theory', in D. Crowley and D. Mitchell (eds), *Communication Theory Today*, Cambridge: Polity.

Meyrowitz, J. (1997). 'Shifting Worlds of Strangers: Medium Theory and Changes in "Them versus Us,"' *Sociological Inquiry*, Vol. 67, No. 1: 59–71.

Meyrowitz, J. (1999). 'Understandings of Media', *ETC: A Review of General Semantics*, Vol. 56, No. 1: 44–53.

Mitchell, W.J. (1996). *City of Bits: Space, Place and the Infobahn*, Cambridge, MA: MIT Press.

Morelli, N. (2001). 'The Space of Telework: Physical and Virtual Configurations for Remote Work', in D. Holmes (ed.), *Virtual Globalization: Virtual Spaces/Tourist Spaces*, London: Routledge.

Morley, D. (1980). *The 'Nationwide' Audience: Structure and Decoding*, London: British Film Institute.

Morley, D. (1986). *Family Television: Cultural Power and Domestic Leisure*, London: Comedia.

Morley, D. (1992). *Television Audiences and Cultural Studies*, London: Routledge.

Morse, M. (1998). *Virtualities: Television, Media Art, and Cyberculture*, Bloomington: Indiana University Press.

Morris, M. and Ogan, C. (1996). 'The Internet as Mass Medium', *Journal of Communication*, Vol. 46, No. 1: 39–50.

Moursund, D. (1996). 'How Long Is a Cyberspace Year?', *Leading and Learning with Technology*, Vol. 24, No. 1: 4–5.

Mullan, B. (1997). *Consuming Television: Television and Its Audience*, Oxford: Blackwell.

Nancy, J.-L. (1991). *The Inoperative Community*, trans. P. Conor et al., Minneapolis: University of Minnesota Press.

Negroponte, N. (1995). *Being Digital*, New York: Knopf.

Negt, O. and Kluge, A. (1993). *Public Sphere and Experience: Toward an Analysis of the Bourgeois and Proletarian Public Sphere*, trans. P. Labanyi, J.O. Daniel and A. Oksiloff, Minneapolis: University of Minnesota Press.

Nelson, R. (1997). *TV Drama in Transition: Forms, Values and Cultural Change*, Houndmills: Macmillan.

Newcomb, H. (ed.) (2000). *Television: The Critical View* (6th edn), New York: Oxford University Press.

Newcomb, T. (1953). 'An Approach to the Study of Communication Acts', *Psychological Review*, 60, 393–340.

Nguyen, D.T. and Alexander, J. (1996). 'The Coming of Cyberspacetime and the End of the Polity', in R. Shields (ed.), *Cultures of the Internet: Virtual Spaces, Real Histories, Living Bodies*, London: Sage.

Nie, N. and Erdring, R. (2000). 'The Internet and Society Study' conducted by the Institute for the Quantitative Study of Society, available at http://www.stanford.edu/group/siqss/

Nightingale, V. (1996). *Studying Audiences: The Shock of the Real*, London: Routledge.

Nisbet, R. (1970). *The Social Bond*, New York: Knopf.

Noble, D. (1997). *The Religion of Technology: The Divinity of Man and the Spirit of Invention*, New York: Knopf.

Norris, C. (1982). *Deconstruction: Theory and Practice*, London: Methuen.

Ong, W. (1982). *Orality and Literacy: The Technologization of the World*, London: Methuen.

Ostwald, M. (1997). 'Virtual Urban Futures', in D. Holmes (ed.), *Virtual Politics: Identity and Community in Cyberspace*, London: Sage.

O'Sullivan, T., Hartley, J., Saunders, D., Montgomery, M. and Fiske, J. (1994). *Key Concepts in Communication and Cultural Studies* (2nd edn), London: Routledge.

Owen, B.M. (1999). *The Internet Challenge to Television*, Cambridge, MA: Harvard University Press.

Plant, S. (1998). *Zeros and Ones: Digital Women and the New Technoculture*, London: Fourth Estate.

Porter, D. (ed.) (1997). *Internet Culture*, London: Routledge.

Poster, M. (1990). *The Mode of Information: Post-structuralism and Social Context*, Cambridge: Polity.

Poster, M. (1995). *The Second Media Age*, Cambridge: Polity.

Poster, M. (1997). 'Cyberdemocracy: Internet and Public Sphere', in D. Holmes (ed.), *Virtual Politics: Identity and Community in Cyberspace*, London: Sage.

Poster, M. (2000). *What's the Matter with the Internet?* Minneapolis: University of Minnesota Press.

Postman, N. (1993). *Technopoly: The Surrender of Culture to Technology*, New York: Knopf.

Puro, J.-P. (2001). 'Finland: A Mobile Culture', in J.E. Katz and M.A. Aakhus (eds), *Perpetual Contact: Mobile Communication, Private Talk, Public Performance*, Cambridge: Cambridge University Press.

Rafaeli, S. (1988). 'Interactivity: From Media to Communication', in R.P. Hawkins, J.M. Wiemann and S. Pingree (eds), *Annual Reviews of Communication Research*, Vol. 16: *Advancing Communication Science: Merging Mass and Interpersonal Process*, Beverly Hills, CA: Sage.

Rafaeli, S. and Sudweeks, F. (1997). 'Networked Interactivity', *Journal of Computer-Mediated Communication*, Vol. 2, No. 4. Available at: http://www.ascusc.org/jcmc/vol2/issue4/rafaeli.sudweeks.html

Real, M. (1984). *Supermedia*, London: Sage.

Reeves, B. and Nass, C. (1996). *The Media Equation: How People Treat Computers, Television, and New Media Like Real People and Places*, Cambridge: Cambridge University Press.

Regis, E. (1990). *Great Mambo Chicken and the Transhuman Condition: Science Slightly over the Edge*, London: Viking Press.

Rheingold, H. (1994). *The Virtual Community: Finding Connection in a Computerized World*, London: Secker and Warburg.

Robins, K. (1995). 'Cyberspace and the World We Live in', in M. Featherstone and R. Burrows (eds), *Cyberspace, Cyberbodies, Cyberpunk: Cultures of Technological Embodiment*, London: Sage.

Robins, K. and Webster, F. (1999). *Times of the Technoculture: From the Information Society to the Virtual Life*, London: Routledge.

Rose, N. (1996). 'The Death of the Social? Re-figuring the Territory of Government', *Economy and Society*, Vol. 25, No. 3: 327–56.

Rosen, R. (1986). 'Search for Yesterday', in T. Gitlin (ed.), *Watching Television: A Pantheon Guide to Popular Culture*, New York: Pantheon.

Rosenberg, M.J. (2001). *E-learning: Strategies for Delivering Knowledge in the Digital Age*, New York: McGraw-Hill.

Rothenbuhler, E.W. (1998). *Ritual Communication: From Everyday Conversation to Mediated Ceremony*, London: Sage.

Russell, G. and Holmes, D. (1996). 'Electronic Nomads? Implications of Trends in Adolescents' Use of Communication and Information Technology', *Australian Journal of Educational Technology*, Vol. 12, No. 2: 130–44.

Saussure, F. de (1992). *Course in General Linguistics*, trans. R. Harris, La Salle, IL: Open Court Classics.

Schultz, T. (2000). 'Mass Media and the Concept of Interactivity: An Exploratory Study of Online Forums and Reader Email', *Media, Culture & Society*, Vol. 22: 205–21.

Schwoch J. and White, M. (1992). 'Learning the Electronic Life', in J. Schwoch, M. White and S. Reilly, *Media Knowledge: Readings in Popular Culture*, Albany: State University of New York Press.

Seiter, E. (1999). *Television and New Media Audiences*, Oxford: Oxford University Press.

Selby, K. and Cowdery, R. (1995). *How to Study Television*, Houndmills: Macmillan.

Sennett, R. (1978). *The Fall of Public Man*, New York: Knopf.

Shannon, C. and Weaver, W. (1949). *The Mathematical Theory of Communication*, Urbana, IL: University of Illinois Press.

Sharp, G. (1983). 'Intellectuals in Transition', *Arena Journal* (New Series), No. 65: 84–95.

Sharp, G. (1985). 'Constitutive Abstraction and Social Practice', *Arena*, 70: 48–82.

Sharp, G. (1993). 'Extended Forms of the Social: Technological Mediation and Self-Formation', *Arena Journal* (New Series), No. 1: 221–37.

Shea, V. (1994). *Netiquette*, San Francisco: Albion Books.

Shields, R. (1994). 'Fancy Footwork: Walter Benjamin's Notes on Flânerie', in Keith Tester (ed.), *The Flâneur*, London: Routledge.

Shields, S. (ed.) (1996). *Cultures of the Internet: Virtual Spaces, Real Histories, Living Bodies*, London: Sage.

Shils, E. (1957). 'Daydreams and Nightmares: Reflections on the Criticism of Mass Culture', *The Sewanee Review*, Vol. 65, No. 4.

Shils, E. and Young, M. (1953). 'The Meaning of the Coronation', *Sociological Review*, December: 63–81.

Silver, D. (2000). 'Looking Backwards, Looking Forward: Cyberculture Studies, 1990–2000', in D. Gauntlett (ed.), *Web.Studies: Rewiring Media Studies for the Digital Age*, Oxford: Oxford University Press.

Silverstone, R. (1999). *Why Study the Media?* London: Sage.

Simmel, G. (1971). 'The Metropolis and Mental Life', in D.N. Levine (ed.), *On Individuality and Social Forms*, Chicago: University of Chicago Press.

Skinner, D. (2000). 'McLuhan's World and Ours', *The Public Interest*, Winter: 52–64.

Skog, B. (2001). 'Mobiles and the Norwegian Teen: Identity, Gender and Class', in J.E. Katz and M.A. Aakhus (eds), *Perpetual Contact: Mobile Communication, Private Talk, Public Performance*, Cambridge: Cambridge University Press.

Slater, P. (1971). *The Pursuit of Loneliness: American Culture at the Breaking Point*, London: Allen Lane.

Slevin, J. (2000). *The Internet and Society*, Cambridge: Polity.

Slouka, M. (1996). *War of the Worlds: The Assault on Reality*, London: Abacus.

Smart, B. (1992). *Modern Conditions: Postmodern Controversies*, London: Routledge.

Smith, C. (1997). 'Casting the Net: Surveying an Internet Population', *Journal of Computer-Mediated Communication*, Vol. 3, No. 1, available at http://www.ascusc.org/jcmc/Vol3/issue1/smith.html

Smith, M. (1995). 'Voices from the Well: The Logic of the Virtual Commons', Ph.D. dissertation, University of California, available at http://www.netscan.sscnet.ucla.edu/soc/csoc/papers/voices/Voices.htm

Smith, M. (1999). 'Invisible Crowds in Cyberspace: Mapping the Social Structure of Usenet', in M.A. Smith and P. Kollock (eds), *Communities in Cyberspace*, London: Routledge.

Smith, M. and Kollock, P. (eds) (1999). *Communities in Cyberspace*, London: Routledge.

Smythe, D.W. (1981). 'On the Audience Commodity and Its Work', in *Dependency Road: Communications, Capitalism, Consciousness and Canada*, Norwood, NJ: Ablex.

Snyder, R.W. (1994). 'The Vaudeville Circuit: A Prehistory of the Mass Audience', in J.S. Ettema and D.C. Whitney (eds), *Audiencemaking: How the Media Create the Audience*, London: Sage.

Sobchack, V. (1996). 'Democratic Franchise and the Electronic Frontier', in Z. Sardar and J. Ravetz (eds), *Cyberfutures: Culture and Politics on the Information Superhighway*, London: Pluto Press.

Sohn-Rethel, A. (1979). *Intellectual and Manual Labour: A Critique of Epistemology*, London: Macmillan.

Soja, E. (1996). *Thirdspace: Journeys into Los Angeles and Other Real and Imagined Places*, Cambridge, MA: Blackwell.

Spears, R. and Lea, M. (1994). 'Panacea or Panopticon? The Hidden Power of Computer-Mediated Communication', in *Communication Research*, Vol. 21, No. 4: 427–59.

Springer, C. (2000). 'Digital Rage', in D. Bell and B. Kennedy (eds), *The Cybercultures Reader*, London: Routledge.

Sprinker, M. (1987). *Imaginary Relations: Aesthetics and Ideology in the Theory of Historical Materialism*, London: Verso.

Stearn, G. (ed.) (1968). *McLuhan: Hot and Cool*, Harmondsworth: Penguin.

Steemers, J. (2001). 'Broadcasting is Dead. Long Live Digital Choice', in H. Mackay and T. O'Sullivan (eds), *The Media Reader: Continuity and Transformation*, London: Sage.

Stevenson, N. (1995). *Understanding Media Cultures: Social Theory and Mass Communication*, London: Sage Publications.

Stoll, C. (1995). *Silicon Snake Oil*, London: Pan.

Stone, A.R. (1991). 'Will the Real Body Please Stand Up?: Boundary Stories about Virtual Cultures', in M. Benedikt (ed.), *Cyberspace: First Steps*, Cambridge, MA: MIT Press.

Stratton, J. (1997). 'Cyberspace and the Globalization of Culture', in D. Porter, (ed.), *Internet Culture*, London: Routledge.

Swyngedouw, E. (1993). 'Communication, Mobility and the Struggle for Power over Space', in G. Giannopoulos and A. Gillespie (eds), *Transport and Communications in the New Europe*, London: Belhaven.

Taylor, T. and Ward, I. (eds) (1998). *Literacy Theory in the Age of the Internet*, New York: Columbia University Press.

Terranova, T. (2001). 'Demonstrating the Globe: Virtual and Real Action in the Network Society', in D. Holmes (ed.), *Virtual Globalization: Virtual Spaces/Tourist Spaces*, London: Routledge.

Tester, K. (1996). *The Flâneur*, Cambridge: Cambridge University Press.

Thompson, J.B. (1994). 'Social Theory and the Media', in D. Crowley and D. Mitchell (eds), *Communication Theory Today*, Cambridge: Polity.

Thompson, J.B. (1995). *The Media and Modernity: A Social Theory of the Media*, Stanford, CA: Stanford University Press.

Thompson, J.B. (1995). *The Media and Modernity*, Cambridge: Polity.

Tingle, L. (2002) 'TV Revolution to Create more Free Channels', in the *Sydney Moving Hearald*, 23 April, p. 1.

Tocqueville, A. de (1990). *Democracy in America*, Vol. 2, ed. P. Bradley, New York: Vintage Books.

Toffler, A. (1980). *The Third Wave*, London: Pan.

Tofts, D. (1997). *Memory Trade: A Prehistory of Cyberculture*, North Ryde: 21C Publications.

Tönnies, F. (1955). *Community and Association*, London: Routledge and Kegan Paul.

Touraine, A. (1988). *The Return of the Actor*, Minneapolis: University of Minnesota Press.

Touraine, A. (1998). 'Sociology without Society', *Current Sociology*, Vol. 46, No. 2: 119–45.

Turkle, S. (1984). *The Second Self*, New York: Simon and Schuster.

Turkle, S. (1995). *Life on the Screen: Identity in the Age of the Internet*, New York: Simon and Schuster.

Turow, J. and Kavanaugh, A. (eds) (2003). *The Wired Homestead*, Cambridge, MA: MIT Press.

Urry, J. (2002). 'Global Complexities', Paper presented at the International Sociological Association World Congress of Sociology, Brisbane, Australia, 7–13 July 2002.

Urry, J. (2003). *Global Complexity*, Malden, MA: Polity.

Urry, J. and Rojek, C. (eds) (1997). *Touring Cultures: Transformations of Travel and Theory*, London: Routledge.

Van Dijk, J. (1999). *The Network Society: Social Aspects of New Media*, London: Sage.

Vidich, A.J. and Bensman, J. (1960). *Small Town in Mass Society*, New York: Anchor Books.

Virilio, P. (1997). 'The Overexposed City', in N. Leach (ed.), *Rethinking Architecture*, London: Routledge.

Virilio, P. (1998). 'Is the Author Dead? An Interview with Paul Virilio', in J. Der Derian (ed.), *The Virilio Reader*, Oxford: Blackwell.

Virilio, P. (2000). *The Information Bomb*, London: Verso.

Virilio, P. (2001). 'Speed and Information: Cyberspace Alarm!', in D. Trend (ed.), *Reading Digital Culture*, Oxford: Blackwell.

Wardrip-Fruin, N. and Montfort, N. (eds) (2003). *The New Media Reader*, Cambridge, MA: MIT Press.

Wark, M. (1994). *Virtual Geography*, Bloomington: Indiana University Press.

Wark, M. (1999). *Celebrities, Culture and Cyberspace: The Light on the Hill in a Postmodern World*, Sydney: Pluto Press.

Wark, M. (2000). 'Whatcha Doin', Marshall McLuhan?', *Media International Australia*, No. 94: 89–96.

Wasserman, I.M. (1984). 'Imitation and Suicide: A Re-examination of the Werther Effect', *American Sociological Review*, No. 49: 427–36.

Watts, D. (2000). 'The Internet, the Small World, and the Nature of Distance', Santa Fe Institute, available at http: aries.mos.org/internet/essay.html

Weigel, Van B. (2002). *Deep Learning for a Digital Age: Technology's Untapped Potential to Enrich Higher Education*, San Francisco: Jossey-Bass.

Wellman, B. (ed.) (1999). *Networks in the Global Village: Life in Contemporary Communities*, Boulder, CO: Westview.

Wellman, B and Gulia, M. (1999). 'Virtual Communities as Communities: Net Surfers Don't Ride Alone', in M. Smith and P. Kollock (eds), *Communities in Cyberspace*, London: Routledge.

Wertheim, M. (1999). *The Pearly Gates of Cyberspace: A History of Space from Dante to the Internet*, Sydney: Doubleday.

Whitaker, R. (2000). *The End of Privacy: How Total Surveillance Is Becoming a Reality*, Melbourne: Scribe Publications.

Whittle, D. (1996). *Cyberspace: The Human Dimension*, New York: W.H. Freeman.

Wiener, N. ([1948] 1961). 'Introduction', in *Cybernetics: or Control and Communication in the Animal and the Machine* (2nd edn), Cambridge, MA; MIT Press.

Wilbur, S. (1997). 'An Archaeology of Cyberspaces: Virtuality, Community, Identity', in D. Porter (ed.), *Internet Culture*, London: Routledge.

Wilhelm, A.G. (2000). *Democracy in the Digital Age: Challenges to Political Life in Cyberspace*, New York: Routledge.

Williams, R. (1961). *Culture and Society: 1780–1950*, Harmondsworth: Penguin.

Williams, R. (1974). *Television: Technology and Cultural Form*, London: Fontana.

Williams, R. (ed.) (1981) *Contact: Human Communication and Its History*, London: Thames and Hudson.

Williams, R. (1983). *Towards 2000*, London: Chatto and Windus/Hogarth.

Williams, R. and Edge, D. (1992). *Social Shaping Reviewed – Research Concepts, Findings, Programmers and Centres in the UK*, Edinburgh: PICT Working Papers No. 41.

Willson, M. (1997). 'Community in the Abstract', in D. Holmes (ed.), *Virtual Politics: Identity and Community in Cyberspace*, London: Sage.

Winston, B. (1998). *Media, Technology and Society: A History from the Telegraph to the Internet*, London: Routledge.

Wise, J.M. (1997). *Exploring Technology and Social Space*, London: Sage.

Wolf, M.J.P. (2000). *Abstracting Reality: Art, Communication, and Cognition in the Digital Age*, Lanham, MD: University Press of America.

Zettl, H. (1973). *Sight, Sound and Motion: Applied Media Aesthetics*, Belmont, CA: Wadsworth.

Žižek, S. (1989). *The Sublime Object of Ideology*, London: Verso.

INDEX

Aakhus, M.A., 185–6
  '*Apparatgeist*' and perpetual contact,
    185–6, 192
abstraction, v, 18n, 32, 128, 151, 153–4,
    177, 192
  constitutive, 158–61
  and the Internet, 162–4
Adorno, Theodor, 34
  culture industry, 23–5, 34, 39, 68,
    88–9, 119, 141
  Frankfurt School, 25, 119
  and Max Horkheimer, 24–5, 33, 68,
    88–9, 99, 141
advertising, xii, 34, 37, 43n, 44, 50, 66,
    77, 87, 95–7, 100, 102, 120n, 149, 201,
    212–13
  billboards, 105, 201, 215
Agamben, Giorgio, 174
agora, 60, 68, 73, 195–7, 199, 201, 209
  cosmopolitan-*agora*, 196
  cyberspace as electronic *agora*, 60,
    195, 199–200
  of potential assembly, 17
Alexander, J., 72, 117, 134, 165n
Alford, J., 28
Althusser, Louis, xi, 28–31, 34, 110, 118, 141
  Ideological State Apparatus, 28
  ideology-in-general, xi, 30, 31, 141
  ideology-in-particular, 30, 141
  interpellation, xi, 29–30, 34, 88,
    110, 141
American Association for the
    Advancement of Science's Program
    in Scientific Freedom, Responsibility
    and Law (AAAS), 150–1
Anderson, Benedict, 107, 194
  imagined communities, 80, 112,
    175, 222–3n
Andrejevic, Mark, 219
  Anna Voog, 220–1
  personal web pages, 220–1
  'the work of being watched', 220
Ang, Ien, 42n, 112, 120n
Aoki, Kumiko, 204–5
  three domains of virtual community, 204
Apple, 98

architectures, 138, 143
  broadcast, ix, 9, 13, 20, 53, 66, 95, 99, 110,
    133, 146–8, 206, 220, 222
  communication, 18n, 130, 206
  media, 164, 223n
  'media' walls, 53
  network, ix, 9, 13, 20, 48, 66, 93, 95, 97,
    100, 110, 133, 194, 202, 222
  social, 83, 89, 99, 154, 165, 169
  urban, 168
Arena, 158–60, 164
ARPANET, 47
AT&T, 56
'attention economy', 104
audience communities, 123, 176,
    194, 210–11
  metonymous identification, 214–21
  rituals of, 214–21
audience, 8, 21, 23, 25, 34, 36, 40, 43n, 50,
    58, 59, 84–6, 89–90, 95, 103, 107–13,
    118–19, 122–4, 127, 131, 133–4, 139,
    142–4, 146–7, 149, 154, 161, 164, 176,
    188, 200, 207, 209, 212, 221
  activity, 30, 40, 42n, 50, 58, 70, 112, 143,
    145, 148, 155, 156, 171, 174
  audience studies, 42n, 58, 101, 111–12,
    118–19, 143
  friends, 144, 148, 210
  hot and cool media, 70–1
  para-social interaction, 144, 148, 153
  passive, 9, 18n, 22, 40–1, 70, 93, 143, 200,
    201, 225n
  synchronous and asynchronous, 96,
    103–6, 109, 139
  see also McLuhan, hot and cool mediums
Aurigi, Alex, 120n
Austin, J.L., 108
Australian Consumers Association, 97
Automatic Teller Machine (ATM), 1, 157
avatar, 62–3, 141–3, 150–1, 190, 198,
    202, 224n

bandwidth, 13, 45, 46, 48, 65, 72, 77, 86, 97,
    106, 210, 221
Barlow, John Perry, 47, 48
  Barlovian cyberspace, 47, 62

Barney, Darin, 191
Barr, Trevor, 8, 79, 95
Baudelaire, Charles, *flânerie*, 197–8
Baudrillard, Jean, xi, 5, 10, 27, 31, 36–8, 39,
    43n, 51, 69, 106, 107, 118, 120n, 133, 143
  hyperreal, 31, 38, 43n, 107
  obscene, 105, 107
  simulacrum, 36, 38, 39, 43n, 100, 118,
    119, 128, 135, 207
Baym, Nancy, 87, 97
  audience community and network
    community, 176, 194, 210
  contexts of CMC, 63–4, 67
Becker, Barbara, 76–7, 79, 81
  'partial publics', 76, 79, 81
behaviourism, x, 5, 18n, 23, 42, 112, 194
Bell, Daniel, 7, 22, 170
  'end of ideology', 22
  multi-stranded community, 170–1
Benedikt, M., Judeo-Christian narratives
    of cyberspace, 190
Benjamin, Walter, *flânerie*, 197–9
Bennett, Tony, 21–2, 42n
Big Brother, 219–20
Bolter, J.D., and Grusin, R., 19n, 38, 43n,
    66, 130
  hypermediacy, 130
  remediation, 19n, 43n
Boorstin, Daniel J., 31–3, 43n
  'homogenisation of experience', 33
Bott, Elizabeth, 99, 166n
Bourdieu, Pierre, 108
Brecht, Bertolt, 19n
broadcast, 11–14, 17, 20–43, 44–5, 49–53, 55,
    57, 59, 64–7, 70, 72–3, 78–80, 82n,
    83–121, 122–3, 132–3, 135, 137–41, 142,
    143, 144–9, 150, 154, 155, 161, 164, 167,
    176, 177, 194, 195, 200, 206–21
  architecture, ix, 9, 13, 20, 53, 66, 95, 99,
    110, 133, 146–8, 206, 220, 222
  authority of, 215, 216
  'The Broadcast Era', 85, 102
  broadcast event, 103–6, 109, 111–12
  as constitutive of media 'mass', 102
  convergence thesis, xi
  dependence on, 3
  interactivity, ix, xi, 10, 12, 50, 84–6, 97–8,
    100, 115, 122, 148
  maintenance of social order, 207
  mediation of visibility, 208–9
  mutually constitutive with network
    integration, 83–6
  network mediums parasitic on
    broadcast, xi, 12, 52, 79, 86–7, 90,
    95, 104, 105

broadcast, *cont.*
  public sphere, 80, 102
  reciprocity without interaction, 144–9,
    161, 195, 211, 217
  rethinking of, 101–19
  sociological approach to, ix-x
broadcast communities, x, xiii, 122,
    206–21, 222
  in news, 215–16
  reality TV, 219–20
  rituals of, 214–20
  symbolic inequality of, 34, 105, 210–14
  talk shows, 217–19
Brosnan, M., technophobia, 183, 188
Buzzard, K.S.F., 18n, 96, 121n

cable, 1, 8, 66, 94, 97, 102, 109, 110, 111,
    121n, 149, 215
Caldwell, J., 120n
Calhoun, Craig, xii, 43n, 154, 160, 164,
    166n, 178
  efficiency of communication, 202
  phenomenological levels of
    socialization, 155–8
capitalism, 4, 6, 9, 24–8, 31, 32, 34, 74, 77,
    82n, 88, 89, 91, 173–4, 207, 221
Carey, James, xii, 6, 15, 39, 46, 117, 215, 219
  anthropological or ritual approach, xii,
    133–135, 208
  divertissement, 133
  status conflict in communication, 219
  uses and gratification model, 134
Carkeek, Freya, 148, 159, 160
cartoon, 70, 71, 120n
Castells, Manuel, 8, 43n, 48, 85, 204–5
  interactive society, 8
CBS, 21, 120n
celebrity, 24, 35, 52, 82n, 212–14, 218–19
  as the last *flâneur*, 214
  Celebdaq and Hollywood SX, 212
  deaths of, 109, 111, 214, 224n
  and non-celebrity, 34–5, 105, 153, 211,
    212–13, 217–18, 219, 220–1, 224n
  and scandal, 214
  and symbolic inequalities, 34,
    105, 210–14
  WorldLive.com, 213
Centre for Contemporary Cultural
    Studies (CCCS), 17n
Certeau, Michel de, 117,
    176, 177
  'space is a practiced place', 176
Chan-Olmsted, S., 96
Chapman, Mark David, 225n
Chesher, Chris, 55, 82n

Cheung, C., 220–1
cinema, 13, 21, 22, 32, 40, 45, 68, 70–2, 86,
    97, 98, 106, 114, 120n, 201, 224n
citizenship, 10, 22, 63, 74, 80, 165n
city, 16, 67, 69, 78, 88, 92, 121n, 156,
    197–201, 223n, 224n
coffee houses, 60–1, 73–4, 77, 82n, 196
communication and information
    technologies (CITs), 2–3, 11–12, 14,
    41, 63
    producing 'new' social relationships,
        19n, 143
'communication as culture', xii
communication environments, x, 12,
    13, 81, 87
community, xii–xiii, 5, 6, 10, 16, 20, 23, 46,
    52, 63, 74, 108, 118, 121n, 123, 133, 148,
    157, 161, 165, 166n, 167–222, 222n,
    223n, 224n
    as a legitimating narrative, 167
    of belief, 176
    citizenship, 10, 22, 63, 74,
        80, 165n
    classical theories of, 167–71
    community as practice, 122, 174–7
    community as recognition, 122
    electronic, 74
    global communities, 172–3, 174–7
    and 'imagined communities', 80, 112,
        175, 222–23n
    imagined universal community,
        129, 189–93
    and 'impact assessment'
        research, 167
    'international' community, 173
    media-constituted, x
    micro-communities, 170, 175
    'miniaturization' of, 169
    network, 176, 188–9, 194, 196
    nostalgia of, 16, 202
    Olympic community, 173
    on-line communities, 87, 101, 150,
        176, 204
    over-integration and over-regulation,
        169–70
    pseudo-communities, 204
    renewal of, xii, 120n, 192
    telecommunity, xii, 17n, 111,
        122, 167–222
    virtual communities, x, xii–xiii, 9, 48,
        54, 62, 63, 68, 78, 80, 99, 100, 103,
        117, 119, 122, 123, 149, 164, 170, 173,
        188, 192, 194–7, 198, 201–6, 222,
        224n, 225n
'complexity theory', 19n, 57

computer-mediated communication
    (CMC), ix–xi, 17n, 47, 54–5, 57, 59,
    67, 87, 90, 103, 117, 118, 119, 123, 137,
    150, 151, 161, 166n, 178, 196, 200, 202,
    205, 211
    as anti-hierarchical, 61
    contexts of, 63–4
    as cyberspace, 60–1, 119
    electronic democracy, 9–10, 73, 76, 80–1
    extends face-to-face communication, 203
    form of 'socially produced space', 60
    and identity, 62–3
    and the public sphere, 78–9
Constitutive Abstraction, 158–62
content (of communication), 25, 55, 57, 60,
    82n, 86–7, 101, 103–6, 107–8, 111,
    113–16, 118, 126, 133, 134, 135, 145,
    163, 166n, 175–6, 211, 215
    content analysis, 6, 59, 119
    focus of media studies, ix, 4–5
    and ideology, 26, 29–30, 34, 141
    and the medium, 36, 39–40, 43n, 49,
        94, 106, 112, 116, 123
    and the user, 143–4, 220
    versus form, ix–xi, xiii, 5–6, 8, 20, 27,
        43n, 51, 56, 118, 143
convergence perspective, xi–xii, 13, 64–6
    corporate, 65;
    functional, 64
    industry, 64
    medium, 64–5
    of space, 67
    technological, xi, 3–4, 41, 50, 64
    telecommunications, xi, 64
Cooley, C.H., 154–6, 160, 170
Corner, John, 18n, 101
Couldry, Nick, 113, 120n, 165–6n, 225n
    media as mythological 'centre' of social
        life, 210–11
    symbolic inequality of media
        friends, 152–3
    talk shows and the 'really real', 218
    talk shows as ritual, 218
Crowley, David, 41, 42
cultural capital, 34
cultural studies, concerns of, 4, 5, 42n, 43n
    concern for passive audience, 22
    influence on media studies, 4, 117
culture industry, xii, 9, 31, 39, 50, 52, 68,
    85, 88–9, 97, 119, 141, 148–9, 161, 195,
    207, 213–14
    mass media as, 23–5
    and the personal web page, 220
cybercafé, 78, 223n
cybergeist, 192

cybernetics, 55–7, 192
cybersex, 161
cybersociety, xi, 44–82
cyberspace, xiii, 1, 2, 4, 9, 11, 18n, 42, 44–7,
    49, 51–2, 54–5, 57, 60–2, 72, 82n, 84, 89,
    101, 117, 119, 133, 138, 140, 150, 175,
    177, 186, 189–93, 195, 196, 203, 222, 223n
  as electronic *agora*, 60, 195, 197, 199–200
  as new public sphere, 72–5, 120n
  as routine, 177
  as revelation, 190
  and 'cyberbrats', 187
  and global telecommunications, 46
  and the Internet, 4, 46–7, 48, 50, 114
  Judaeo-Christian narratives of, 190
  and virtual *flânerie*, 199–201
  and virtual reality, ix, 44–6, 49, 122, 201
cyber-terrorism, 11
cyber-utopia, 52, 75, 83, 115, 120n, 191, 194

Dallas Smythe, 43n
datacasting, 84, 103–5, 106, 120n
Dayan, D., 107, 208, 216
Debord, Guy, 5, 27, 53, 68, 90
  spectacle, 31–4, 89, 118
deconstruction, 5, 6
Deleuze, Gilles, rhizome, 10
democracy, 23, 43n, 73, 76, 77, 82n, 134,
    207, 208
  and cyberspace, 24, 72–5, 80, 191
  and interaction, 80–1
democratization, 9–11, 23, 84, 207, 220
Dempsey, Ken, 175, 194
Derry, Jacques, 6, 20, 51, 121n, 129, 131,
    133, 165n, 223n
  dissemination, 126–8
  'hermeneutic deciphering', 126, 131
  logocentrism, 6, 11, 75, 123–4, 127–9,
    130–2, 135, 138, 141, 146, 166n
  phonocentrism, 6, 124, 129, 132,
    135–6, 139
  polysemia, 126–8
  writing-as-language, 131
Dery, Mark, 100
  'escape velocity', 187
dialogicity, 40, 77–9, 136, 137, 146–7, 164,
    203, 205
digital, 'digital age', 10
  digital divide, 58, 187
  'digital nation', 73
  technology, 2, 7, 8, 19n, 45, 49, 60, 64,
    65–6, 82n, 100, 103, 108, 114, 115,
    130, 183–184
digitalization, 65–6, 164
disembedding, 162–3

disembodiment, 36, 100, 157, 159–161, 178,
    200, 207
  disembodied communities, 194, 225n
'disintermediation', 137–8
dot.com stocks, 96–7
Durkheim, Emile, 52, 59, 121n, 133, 152–3,
    154, 189
  community, 167–9, 175–6, 182, 194
  conscience collective, 110, 167–70, 175
  cult of the individual, 29, 110
  'organic solidarity', 168
DVD, 65–6, 100

e-commerce, 114
economy, 173
effects analysis, 4, 21, 42n, 56, 58, 82n, 102,
    115, 119, 123, 133, 136, 166n
electrical-analogue time-worlds, 49
electronically based communications, 8
electronically extended relations, 3, 39,
    49, 54, 55
email, 3, 17n, 47, 50, 55, 60, 61, 78, 79, 94,
    97, 104, 116, 120n, 132, 143–4, 145,
    150, 166n, 204, 207
embodiment, xiii, 2, 12, 14, 16, 36, 42, 47,
    55, 60–3, 67, 69, 73, 78, 80, 84, 90, 92,
    94, 99, 137, 138, 141, 145–146, 154,
    157–8, 160–1, 170, 188, 192, 196, 200,
    203, 207, 210, 217, 223n, 224n, 225n
emoticons, 16, 55, 82n, 161
Endemol Corporation, 219
Erdring, R., 90, 194, 224n
European traditions in media studies, 4

face-to-face interactions, x, xii, 2, 8, 11–12,
    14, 15–17, 17n, 49, 54, 63, 71–2, 78, 81n,
    85, 87, 92, 94, 99–100, 108, 111, 114,
    116, 118, 119n, 123n, 132, 135, 136–9,
    144–6, 148–51, 154–6, 158–9, 161, 164,
    166n, 178–80, 194–5, 197–9, 204–5, 207,
    211, 221–2, 224n, 225n
  extended by CMC, 54–5, 63, 106, 118, 203
fandom, 87, 112, 153, 211, 212–13, 214
  on-line fan clubs, 87
Featherstone, Mike, the *flâneur*, 197, 199–201
  MTV, 201
Feenberg, A., 140
Felski, Rita, 75
Fidonet, 47
Fiore, Quentin and Marshall McLuhan,
    41, 72, 99, 103, 118, 121n,186
first media age, 6, 34, 95, 120n, 140
  and the second media age, ix, 4, 7–11,
    12, 17, 43n, 44, 50, 52, 67, 69, 71,
    82n, 83–91, 97, 110, 114, 140, 194, 204

Fiske, John, 57, 82n, 120n
  'bardic' function of news, 110, 121n, 215
the *flâneur*, 197–198, 210, 223–4n
  and celebrity, 214
  defined as consumer, 199
  virtual, 199–201, 210, 223–4n
flânerie, 170, 197–203, 207, 210, 224n
Flew, Terry, 64, 65, 137
Flitterman-Lewis, S., 106
form, versus content, ix–xi, xiii, 5–6, 8, 20,
    27, 43n, 51, 56, 118, 143
Forrest Gump, 34–5
Foster, Derek, 190, 194, 204, 205
Foucault, Michel, 31–3, 43n, 153
  disciplinary society, 33
  'governmentality', 172, 174
Frankfurt School, 25, 119
Fraser, Nancy, 75
freeway, xii, 67, 68–9, 99
Friedberg, Ann, 199
ftp, 79
Fukuyama, Francis, 99, 169

Gates, Bill, 7
Gauntlett, D., 7, 18n, 96, 104, 105, 180, 224n
generation gap, 19n, 63
genre, 14, 24, 34–5, 36, 37, 49, 87, 100, 105,
    107, 112, 113, 133, 134, 139, 147, 153,
    201, 210, 211, 212, 213, 214–19, 224n
Gerbner, George, 57–8, 60, 119
  access, and availability, 58
  vertical dimension of communication, 58
Gibson, William, 45, 47, 189, 196
Giddens, Anthony, xii, 21, 154, 162–3
  time-space distanciation, 162–3
Gilder, George, 7, 9, 10, 14, 52, 84
Gitlin, Todd, 17n, 42n, 76
  'public sphericules', 75–6, 81
global communities, 129, 172–3, 189–93
  global citizenship, 63, 80
  of practice, 174–7
globalization, 107, 163, 167, 168,
    172, 173–4
Goffman, Erving, 154
gopher, 79
Gore, Al, 7, 18n, 74
Graham, S., 43n, 67, 120n
Gramsci, Antonio, 28
  hegemony, 28, 40, 43n, 52, 101, 153
Grusin, R. and Bolter, J.D., 19n, 38, 43n,
    66, 130
  hypermediacy, 130
  remediation, 19n, 43n
Guattari, Félix, 16
Gulia, M., 17n, 202, 204

Habermas, Jurgen, 209
  public sphere, 42n, 72–81
Hall, Stuart, 17, 26, 117
  'American Dream Sociology', 22–3
  encoding/decoding, 17n, 112
Hanks, W.F., 176
Hartley, John, 18n, 22, 42n, 73, 77, 120n
  'bardic' function of news, 110, 121n, 215
Hawisher, G.E., 74, 75, 82n
Healy, Dave, 54
Hegel, G.W.F., 107, 127, 185, 223n
Heidegger, Martin, 140, 181–2, 187, 191
Heilig, Morton, 81n
Herbert, T.E., 46
Hill, A., 180
Hills, Mathew, 87, 112
Hirst, P., 30, 31, 80
historicism, ix, xi, xii, 7–11, 39, 64, 65, 81,
    83–6, 97, 129, 145, 192, 206, 221
  problems with historical typology, 11–15
Hobbes Internet Timeline, 81
Hollywood, 24, 104, 105, 213
Horkheimer, Max, and Theodor Adorno,
    24–5, 33, 68, 88–9, 99, 141
Horrocks, Christopher, 72, 114, 115,
    121n, 143
Horton and Wohl, *The Lonesome Gal*, 212
  para-social interaction, 144, 148, 152–3,
    166n, 212
HTML, 60
hypodermic model, 58

IBM, 47
ICQ, 50, 60, 61
identity, 5, 8, 36, 76, 92, 123, 127, 139,
    141–4, 149, 150, 165n, 174
  constituted by media environments,
    15, 21, 48, 53, 61–3, 99, 151, 180,
    184, 189, 190, 194, 205–7
  see also avatar
ideology, ix, xi, 4, 5, 18n, 22, 34, 37–9, 43n,
    54, 74, 75, 81n, 101–2, 118–19, 128, 130,
    141, 195n
  'end of ideology', 22
  of interactivity, 18–19n
  media as apparatus of, 25–9
  as a structure of broadcast, 29–31
  see also Althusser
IMAX, 92
'Information Revolution', 19n, 56
information society, 2, 7, 55–6, 173
information theory, x, 55–9, 64, 119
  Shannon, C. and Weaver, W.,
    56, 82n
Inglis, F., 107

Innis, Harold, 5, 38, 41–2, 43n, 51, 70, 85, 98, 117
Institute for the Quantitative Study of Society, 90
institutionally extended relations, 49
instrumental views of communication, xii, 18n, 20, 81n, 130, 132, 137, 138, 139–40, 202–3, 206
Integration, x–xiii, 12, 13, 15–17, 29, 35, 53, 55, 69, 75, 99, 119, 120n
  broadcast, 3, 34, 35, 72, 86, 98, 148–9, 177, 206–10
  community as practice and community as recognition, 122, 174–7
  levels of, x, 122, 148, 151–3, 153–60, 165, 166n, 222, 223n
  network, 3, 8, 72, 98, 141, 149, 177
  and ritual view of communication, 119, 122–3, 133–5, 140
  social integration, 3–4, 6, 16, 21–2, 24–5, 31, 50, 52, 55, 83, 86, 118, 122–3, 132, 145, 148, 152–7, 159, 164, 168, 169–71, 173, 175, 182, 183, 206, 208, 210–11, 222
  and sociality with mediums, 178–80
  versus interaction, 15–17, 55, 122–66
Interaction, x–xiii, 8, 12, 14, 40, 49, 52, 76, 77–9, 84, 86, 90, 99, 114–16, 118, 175–6, 177, 178, 180, 185, 188, 195–9, 202–4, 206, 209–10, 219, 222, 224n, 225n
  approach to media culture, xi, 164, 178
  and broadcast, 52, 53, 73, 86–8, 106, 108, 110, 112–13
  and democracy, 80–1
  face-to-face, *see* face-to-face interactions
  world divided between the 'interacting' and 'interacted', 8
  mediated, 15, 86, 115–16, 136–9, 145, 160
  mediated quasi-, 136–7, 144–8, 160
  para-social, 144, 148, 153, 166n
  practice of, 77, 175
  without reciprocity, 144–9, 161, 195, 211, 217
  technologically extended, 76, 86, 94–5, 98, 137, 159, 222
  and transmission view of communication, xii, 15, 119, 164, 177
  virtual interaction, 61–4, 79, 108, 200, 224n
  without reciprocity, 149–51, 195, 205
interactivity, 9, 10, 12–13, 18n, 40, 53–54, 72, 75, 80, 82, 89, 94, 100, 114–15, 117, 195–7, 200, 203–6, 220, 222, 223n
  activity versus passivity, 14, 15

interactivity, *cont.*
  and broadcast, ix, xi, 10, 12, 50, 84–6, 97–8, 100, 115, 122, 148
  interactive society, 8
  versus connectivity, 206
interface, 14, 60, 63, 64, 91, 187, 188, 223n
  human-technical interface, 2, 180
  Interface Culture, 223n
'Internet and Society' study, 90, 224n
Internet, ix–xiii, 2, 3–5, 7–17, 17n, 18n, 19n, 31, 36, 40, 42, 44–5, 55–7, 61, 63–4, 66–7, 72–3, 76–80, 81n, 82n, 83–90, 92–3, 95–104, 106, 108, 110, 113–15, 118, 120n, 121n, 122, 128, 132, 141–3, 145, 155, 157, 166n, 169, 175, 177, 183, 189, 190, 193, 201–4, 207, 209, 214, 219, 220–2, 223n, 224n
  abstraction and, *see* abstraction
  anarchy of, 8
  ARPANET, 47
  attraction of Internet communication, 48–50
  bulletin board, 17n, 47, 78, 94, 100–1, 104, 196
  and cyberspace, 4, 46–7, 48, 50, 114
  digital divide, 58, 187
  electronic frontier, 9, 48, 65, 202
  email, *see* email
  emancipation from broadcast media, 9, 44, 48, 50–4, 84
  emoticons, 16, 55, 82n, 161
  Fidonet, 47
  ICQ, 50, 60, 61
  interaction without reciprocity, *see* Interaction without reciprocity
  'Internet' age, 10, 49, 54, 80
  Internet community, 194–7, 201, 222
  Internet datacasting, 84, 103–5, 106, 120n
  Internet Relay Chat (IRC), 17n, 61, 79, 166n
  'Internet Revolution', 1, 51
  Internet Service Provider (ISP), 97
  MOOs, 17n, 48, 60–1, 93
  MUDs, 17n, 48, 50, 60–1, 184
  Net flaming, 69, 100
  netiquette, 16, 55, 82n
  netizen, 99, 101, 115, 190
  newsgroup, 17n, 47, 87, 202
  reconstitution of public sphere, 72–5, 120n
  redemptive of interactivity, 9, 10, 44, 54, 84, 89, 120n, 189–90, 203
  sub-media, 12, 47, 48–50, 77, 79, 80, 90, 94, 96, 104, 150–1, 188, 200, 224n
  Usenet, 47, 57, 79, 87

Internet, *cont.*
  and the virtual *flâneur*, 197, 199–201,
    210, 223–4n
  the WELL, 47, 61–2, 132, 225n
  World Wide Web, *see* World Wide
    Web (WWW)
intranet, 47, 196
ISDN, 80
IVF, 158

Jakobson, Roman, 56
James, Paul, 148, 159–60
Jameson, Frederic, 32
JennyCam, 120
Johnson, Steven, 223n
Jones, Steven, 7, 14, 18n, 60, 62, 68, 194,
    200, 204
Jordan, Tim, 38, 47, 50, 61–2, 82n
  avatar, 62
  CMC as anti-hierarchical, 61
  cyberpower, 193
Jowett, Garth, 85–6

Kaplan, N., 74
Kapor, Mitch, 48, 84
Katz, Elihu, 107, 112, 117,119, 133, 208, 216
Katz, Jon, '*Apparatgeist*' and perpetual
    contact, 185–6, 192
  'digital nation', 73
Kelly, Kevin, 73, 190
Kling, Rob, 150
Kluge, Alexander, 75
Knorr Cetina, Karin, 17n,
    117, 181–3
  on Heidegger, 181
  objectualization, 182
  post-social, 182
Knowles, Harry, 104–5
Kroker, Arthur, 19n, 92–3, 115,
    117, 143
  on McLuhan and the 'new universal
    community', 189
Kroker, Marilouise, 92–3, 143

Lacan, Jacques, 6, 165n
Lakoff, George, 56, 203–4
Langer, J., 148, 224n
Lanier, Jaron, 191
Lasn, K., 120n
Lasswell, H., 58–9, 119
Lave, J., 176
Lea, M., 82n, 166n
Lealand, Geoff, 19n
Lennon, John, 214, 224n, 225n
Levinson, Paul, 99, 115

Lévy, Pierre, 7, 9
linguistic perspective on media, x,
    4–6, 18n, 23, 51, 56, 60, 101–2,
    126, 159
  semiolinguistics, 124, 128–9, 165n
Lipset, Seymour, 22–3
'liveness', 96, 106, 107, 120n, 143, 153, 209,
    215, 216–17
Livingstone, S., 7, 224n
local context, 111, 128, 138–9, 141,
    145–6, 162
logos, 11, 124, 126, 130
Los Angeles, 120n
Luhmann, Niklas, v., 144
Lukács, Georg, 27
  'reification', 26, 27, 31, 32, 36, 55, 89
Lyotard, Jean-François, 6, 70
  grand narratives, 11, 128, 173

magazines, 24, 25, 36, 56, 73, 96, 105, 112
malls, xii, 37, 67, 68, 195
  and *flânerie*, 170, 199
  privatization of public space, 3, 91
Marc, David, 4, 102, 109–11, 114, 120n,
    121n, 165n
Martin-Barbero, J., 99, 209, 210
Marvin, S., 43n, 67
Marx, Karl, 26–27, 37, 43n, 191
  commodity fetishism, 26, 37
mass media, 5, 7, 9–10, 13, 20, 21–3, 32,
    34–8, 40, 42n, 51, 58, 76, 80, 88–90, 96,
    99–100, 104–5, 112, 137, 141–2, 144–5,
    157, 164, 166n, 176, 195, 196, 200–1,
    209, 212
  as agent of integration, *see* Integration,
    broadcast
  as apparatus of ideology, 25–9, 29–31
  as a culture industry, *see* culture
    industry, mass media as 'path
    dependence'
mass society, 21–25, 42n, 82n, 136
  'age of the masses', 22
  'massification' of society, 21
mass/elite framework, 21–2, 38
Mattelart, Armand, 54
McCarthy, Anna, 37, 208
McLaughlin, L., 217
McLuhan, Marshall, xi, 5, 8, 38–42, 43n, 51,
    69–72, 82n, 94, 99, 103, 107, 113,
    114–17, 118, 121n, 129, 142, 143–4, 154,
    177, 189, 199, 210, 219
  automation, 40, 69
  cybernation, 40, 69–71
  global village, 39, 74–5, 80, 128–9, 164,
    189, 197

McLuhan, Marshall, *cont.*
Gutenberg or typographic man, 129
hot and cool mediums, 40–1, 70–2,
114, 210
McLuhan Galaxy, 8
media and narcissism, 180–1
'new universal community', 189
rear-view mirrorism, 113, 121n
're-tribalization', 69, 72
sociality with objects, 180–1
'the medium is the message', 38–42, 143
'the medium is the massage', 38, 41
'to go outside is to be alone', 199
virtuality, 114, 143
youth and the 'electric drama', 186
McQuail, Denis, 39
media effects theory, 4, 42n, 56, 82n, 115
see also effects analysis
media environments, xi, 14, 144, 179,
184, 186
media event, 45, 66, 104, 107, 112, 134, 149,
153, 208, 214, 216
as interruption of routine, 107, 216
'contest, conquests, and coronations', 107
Coronation, 107, 207–8
Olympics, 104
Princess Diana, 109, 111, 121n
media friends, 211, 213
media studies, ix–xi, 4–7, 18n, 42n,
43n, 55, 58, 82n, 84, 101, 102, 116–19,
133, 178
focus on content and representation,
ix, 4–5
media stunt, 105
mediation, x, xii, 17n, 42, 52, 89–90, 122,
126, 127, 132, 137–8, 160–2, 164, 173,
202–3, 208, 218
'disintermediation', 137–8
mediation-by-agents, 137
problems with, 138–40
remediation, 19n, 43n
media users, 143
'mediaplace', 100
mediascape, 222;
and sociality, 177
medium theory, 38, 41–2, 51, 56, 101–3,
113–9, 122–3, 178, 181
and individuality, 140–4
Shannon, C. and Weaver, W., 56, 82n
Mehl, Dominique, 165–6n
Mellencamp, Patricia, 107
Meyrowitz, Joshua, xii, 1, 15, 38–42, 43n,
45, 51, 94, 99, 103, 115–17, 121n, 132,
139, 148, 154–5, 162, 164, 178, 211, 213
death of John Lennon, 224n, 225n

Meyrowitz, Joshua, *cont.*
intentionality, 56, 115, 130–2, 139, 141
media friends, 211, 213
medium-as-environment, 103, 115–17,
121n, 141
medium-as-language, 115
medium-as-vessel/conduit, 103,
115–16, 123, 130, 206
'second generation medium
theory', 117–18
sociality with mediums, 178
Microsoft, 184, 223n
MIT Media Lab, 74
Mitchell, David, 41, 42
Mitchell, William J., 67, 183, 199, 200
Morley, Dave, 112–13, 149
Morse, Margaret, 43n, 91, 94, 117, 134
MP3, 66
MTV, 49, 91, 92, 120n, 201
Mullan, Bob, 207–8
broadcast as maintenance of social
order, 207
multi-culturalism, 76
multimedia, 8, 49, 66, 200

Nancy, Jean-Luc, 174
narcissism, 180–1, 184, 204
'narrow-band', 79
narrowcasting, 111
'nation-as-audience', 111
'nation-of-audiences', 111
nation-state, 21, 43n, 75, 80, 81, 171–5,
222n, 223n
Nationwide, 112–13
Negroponte, Nicholas, 7, 9, 10, 18n,
74, 224n
Negt, Oskar, 75
Nelson, Robin, 81–2n
'Netaid', 44, 81n
network, ix–xii, 1–5, 8, 10–15, 16–17, 17n,
20, 34, 38, 42, 44, 46–51, 52, 54, 59,
64–6, 68–9, 70, 72, 74, 78–9, 82n, 83, 85,
89, 90, 95–6, 97–103, 109, 110, 113–14,
117–19, 119n, 121n, 130, 132–4, 138,
141–3, 146, 148–9, 156–7, 160–1, 167,
171, 176, 179–81, 192, 194, 196, 201,
205–6, 210, 224n
community, 176, 188–89, *see also*
virtual community
and democracy, *see* democracy
dependence on, 3, 143, 178, 180
integration, *see* integration, network
as normative medium, 14
parasitic of broadcast medium, xi, 12,
52, 79, 83, 86–7, 90, 104–5

network, *cont.*
  sociality with, *see* dependence
  'network society', 82n, 101, 129, 171
  telecommunity, 188–9
New Criticism, 5
New Media, xii, 8, 13, 17n, 19n, 50, 84, 102,
    114, 117–18, 123, 154–5, 159, 166n,
    185–188
  and digitization, 13
  and ideology, 18n
  'new media age', 7
  New Media historicism, 13
  youth consumption of, 19, 186–7
Newby, H., 170
Newcomb, H., 101, 120n, 121n
Newcomb, T., 58
news, 23, 34, 40, 66, 86, 105, 109, 111,
    112, 120n, 133, 148, 164, 165n, 207,
    215–6, 223n
  as drama, 134–5
  bardic function of, 110
  newsflash, 105
  newstext on-line, 66
Nguyen, D.T., 72, 117
Nie, Norman, 90, 194, 224n
Nightingale, Virginia, 112–13
Nisbet, Robert, 121n
novel, 24, 105, 115
nuclear power, 158

oikos, 77
Olympics, 104
Ong, Walter, 166n
ontology, 15, 36, 39, 193, 200, 216
optical fibre, 2, 49, 64, 66
Ostwald, Michael, 45–6, 67, 196
otherness, 2, 35
Owen, Bruce, 102

packet-switching, 2, 10
pay-per-view media, 97, 108
PBS, 110
PDA, 66
performativity, 38, 55, 96, 108, 109, 111–2,
    121n, 215
  'speech act', 108, 152
  'speech community', 108, 192, 210
personal web-page, 100, 145, 213, 220–1
personalization, 13, 68, 91, 92, 177
person-to-person (P2P), 65
phonetic alphabet, 71
photography, 19n, 70, 71, 73, 82n, 117
Plato, 191, 223n
pluralism, 21, 23, 76, 134, 168
polis, 68, 77

positivism, 5, 18n, 23, 33, 42n, 55, 57–8,
    131, 188
Poster, Mark, 7, 8–9, 11, 15, 51–3, 73, 78, 80,
    82n, 84, 195–6, 221
Postman, N., 17n, 191, 192
post-social, 182–3, 186–8
post-structuralism, 23, 29
presence, 6, 12, 13, 31, 36, 40–1, 61, 81n, 90,
    94, 98, 107, 108, 121n, 124, 127–30,
    132–3, 135–9, 148–9, 152, 159–60, 165n,
    178–9, 203, 221
  self-presence, 124, 127–8, 165n
prime-time, 50
process schools, 57, 82n
Proctor, W.S., 46
propaganda, 21, 58
proto-virtual reality, 67, 81n
public space, 3, 68, 77, 92, 117, 120n,
    199, 224n
public sphere, 9–11, 42n, 72–81, 92, 99, 102,
    154–6, 207, 209
  and CMC, 78–9
  decline of, 73
  democracy and, 23, 72–5, 207
  feminist, 75
  'oppositional' working class, 75
  'post-bourgeois', 75
  reconstitution by Internet, 72–5, 120n
  transformation of, 79
  'public-sphericules', 75–6, 81

radio, 1, 4, 9, 11–13, 17, 19n, 21–2, 25, 34,
    40, 45, 51, 65, 70–1, 85, 87, 89, 96, 102,
    105–6, 108–9, 112, 114–15, 120n, 121n,
    123, 136–7, 144, 147, 196, 208, 212, 223n
  radio on-line, 66
Rafaeli, S., 194, 205–6
rage, 93
  air-, 99
  road-, 69, 99
  street-, 99
  telephone-, 99
RAND Corporation, 10
RealAudio, 115, 144
'real time', 49, 79, 94, 106,
    135, 145
Real, M., 105
reality TV, 14, 37, 100, 153, 212, 219–20
  and the audience, 85, 113, 147
reciprocity, 10, 20, 49, 53, 55, 64, 81n, 85,
    95, 98, 121n, 137, 141, 152, 161, 166n,
    196, 212
  and broadcast, 110, 144–9, 195, 211,
    217, 219
  and Internet, 110, 149–51, 195, 205

recognition, 30, 33–4, 100, 110, 121n, 134, 145, 148–9, 151, 156–7, 164, 208, 211–12
  community and, 98, 122, 168, 174
  face-to-face, 92
  field of, 19n, 22, 36, 111, 151, 214, 217, 218, 219
representation, 31–2, 36–7, 43n, 106, 124, 127, 130, 133, 135, 206–7, 215–16, *see also* 'the image'
  focus in media studies, ix, 4–5
  and identity, 142
Rheingold, Howard, 7, 9, 10, 81n, 84, 97, 132, 188, 195, 224n, 225n
  as a nostalgic communitarian, 16
ritual, xii, 14–15, 17, 20, 55, 60, 87, 111, 118, 131–5, 152–3, 165n, 166n, 177, 183, 188, 207–10, 212, 225n
  audience communities, 214–15, 217, 221–2, 225n
  broadcast communities, 207–10
  versus transmission view, 6, 20, 119, 122–35
  view of communication, x, 6, 17, 119, 140, 147, 177, 222
Rose, Nikolas, 171–4, 182
Rosen, Ruth, 209
Russell, G., 19n, 63

satellite based communications, 2, 13–14, 51, 64, 66, 94
  global positioning system (GPS), 2
'saturation' thesis, 2, 127
  'saturated self', 17n, 155
Saussure, Ferdinand de, 5, 124–5, 165n
  'Copernican revolution' in humanities, 5
Schultz, Tanjev, 12–13, 85, 100–1, 120n, 206, 209, 223n
Schwoch, James, 1–3, 17n
second media age, ix–xi, 1–17, 20, 39, 43n, 44–5, 50, 54–5, 58, 60, 64–5, 67, 69, 72, 77, 80, 83–91, 97, 101, 102, 137, 140, 145–6, 195–6, 201, 220, 222
  as agent of return of *flânerie*, 199–201
  and first media age, ix, 4, 7–11, 12, 17, 43n, 44, 50, 52, 67, 69, 71, 82n, 83–91, 97, 110, 114, 140, 194, 204
  historicism, *see* historicism
  Internet as emancipation from broadcast media, *see* Internet, emancipation
  as orthodoxy, ix, xi, 8, 19n, 20, 50, 65
  're-tribalization', 69, 72
  thesis, ix, xi, 4, 8, 12, 20, 50–4, 55, 64, 70, 82n, 84, 87, 101, 102, 148, 185, 187, 201, 222

second media age, *cont.*
  utopianism, 7, 18–19n, 52, 57, 74–5, 83, 98, 115, 120n, 128, 157, 179, 189, 191–2, 194–5, 202, 224n
Selfe, C.L., 74–75, 82n
semiotics, ix, 5, 11, 23, 51, 82n, 101, 127, 207, 210
  analysis of media, x, xi, 82n, 119
Shannon, C., 55–56, 82n, 119
Sharp, Geoff, 3, 94, 99, 120n, 151, 154, 158–9, 166n, 192
Shields, R., 198
Shils, Edward, 22–3, 207
silicon century, 2
Silverstone, Roger, 3, 18–19n, 180, 207, 223n, 224n
Simmel, G., 121n, 201
Situational/Interactionist perspective, 154–5
Skog, B., 187
Slater, P., 156
Slevin, James, 154, 162–4, 166n
Slouka, Mark, 189, 204
Smith, Marc, 61–2, 64, 82n, 200, 202
SMS, 84, 97, 187, 188
soap opera, 16, 86–7, 111, 119, 148, 209
social architectures, *see* architectures, social
sociality, 152, 172
  'drive for sociality', 195
  with mediums, 177–80
  with objects, 119, 177, 180–3
sociological approach, ix–x, 4, 18n, 22, 25, 42n, 59, 78, 86, 96, 108, 122, 152, 154, 164, 194
Sohn-Rethel, Alfred, 154, 158
Sony Ericsson, 82n
  Walkman, 45
spam, 77, 97, 166n
Spears, R., 82n, 166n
spectacle, xii, 6, 12, 24, 27, 31–6, 38, 43n, 55, 89–90, 110–13, 118–19, 120n, 200, 207–8, 210, 218, 219, 224n
spectatorship, 32, 144, 207–8, 212, 220
speech, 5, 10, 11, 23, 39–40, 43n, 47, 49, 56, 58, 60, 65, 69–72, 74, 95, 105, 108–9, 110, 121n, 126, 129, 135, 152, 181, 192, 196, 205, 209–10
Stenger, Nicole, 191–2
'stimulus' and 'response', 21
Stoll, C., 204, 224n
Stone, A.R., 195–6
Stratton, Jon, 46
subcultures, 8, 43n, 80, 169, 187
subject, 5, 6, 11, 14, 26, 29–30, 33, 34, 42n, 53, 89, 113, 124, 141–3, 170, 174, 178, 188, 193, 205, 211, 221

subject, *cont.*
  as a fiction, 16
  'subject position', 30, 143, 170
suburbanization, 68, 88, 92, 99, 195
  see also city, urbanization, freeway, malls
Sudweeks, F., 194, 205, 206
surveillance, 11, 33, 156, 163, 219–20, 223n

talk show, 24, 165n, 208, 211, 212,
    224n, 225n
  as ritual, 225n
  and audience community, 217–19
  host as intermediary, 217
  and metonymous identification, 217–19
  and the '*really* real', 218, 222–3n
Taylor, T., 205–206
technological determinism, 12, 17n, 81, 85,
    178–9, 186
'techno-social' relations, xi, xiii, 12, 60, 84,
    155, 156, 162, 179–80, 207
'technostructure', 115, 117
telecommunications, 2, 13, 14, 52, 56, 65,
    69, 128–9, 135, 161, 167
  and cyberspace, 46
  convergence, xi, 64
telecommunity, xii, 17n, 111, 122, 167–222
telegraph, 13, 39, 40, 46, 56, 120n, 134
telephone, 1, 2, 12–13, 16–17, 46–7, 48, 51,
    56, 61, 65, 68, 70–2, 79, 85–7, 89, 92,
    97–9, 114–15, 136, 143–4, 165n, 193
television, 1, 2, 4, 8, 9, 11–15, 17, 18n, 19n,
    22, 24, 32–3, 34, 37, 40, 45, 49–52, 58,
    64–8, 70–4, 79, 81n, 82n, 84–7, 89, 92–8,
    100–3, 105–8, 110–12, 114–17, 120n,
    121n, 123, 130, 132, 136–8, 141, 143–8,
    153–5, 159, 165n, 166n, 169, 174, 177,
    180, 185, 188, 196, 200–1, 206, 208–22,
    223n, 224n, 225n
  'TV age', 71, 92, 93
  television studios, 211, 214–15,
    217–18, 225n
telnet, 79
Telstra, 97
Terranova, T., 98
terrorism, 43n
  see also cyber-terrorism
Tester, Keith, 197
  the *flâneur*, 198, 224n
The Ed Sullivan Show, 109
'the image', 5, 15, 31–5, 36–8, 52, 64, 68, 89,
    94–5, 105, 117, 214, 221
Thompson, John B., xii, 21, 33, 85, 136–9,
    141, 144–8, 151–2, 154, 155, 158, 160,
    162, 163–4, 165n, 166n, 208–9, 218
  instrumental/mediation paradigm, 137

Thompson, John B., *cont.*
  'mediated publicness', 34, 76, 223n
time-space, 163–164
  compression, 117
  distanciation, 162–3
  relations, 162–3
Toffler, Alvin, 221–2
Tofts, Darren, 192–3, 223n
  cspace, 192–3
Tönnies, Ferdinand, 167–170
  Gemeinschaft, 168–9, 175, 197, 202, 210
  Gesellschaft, 168–9, 172, 197
  totalization, 127–8
Touraine, Alain, 171–4, 182, 194
  'end of Homo Sociologicus', 171
  decomposition of social norms, 173–4
  'programmed society', 171–2
  society as a technology of managing
    populations, 171
tourism, 24, 170, 207, 224n
transmission view, 6, 42n, 53, 58, 118,
    130–4, 138, 140
  and interaction, xii, 15, 119, 164, 177
  and 'process schools', 57
  versus ritual view, 6, 20, 119, 122–35
'transport' model of communication, *see*
    transmission view
Turkle, Sherry, 2, 7, 49, 52, 54, 80, 141–142,
    144, 183–4
  'age of the Internet', 10, 49, 54, 80
  computer screen as 'second self',
    2, 54, 184
  digital intimacy, 183–4

United Nations (UN), 44, 67, 222n
urbanization, xi, 12, 21, 32, 67–69, 88, 90,
    92, 197
  micro-urbanization, 68
  urban life, x, 3, 53, 54, 67, 68–9, 78, 83,
    91, 120n, 149, 156, 167, 182, 196,
    199, 201, 222
  see also city, freeway, suburbanization,
    malls
Urry, John, 179
Usenet, 47, 57, 79, 87
user perspective, 18n, 59, 143, 180, 223n
'uses and gratification' model, 112, 134

van Dijk, J., 64, 65, 82n
Vaudeville, 119n
video, 49, 50, 66, 78, 100, 103, 115, 200
  games, 74
  video age, 80
  video-cafés, 78
  video-on-demand, 8, 103–4, 108

Virilio, Paul, 17n, 117, 224n
virtual community, x, xii–xiii, 9, 48, 54, 62,
    63, 68, 78, 80, 99, 100, 103, 117, 119,
    122, 123, 149, 164, 173, 188, 192, 194–7,
    198, 201–6, 224n, 225n
  dichotomies of, 179, 203–4
  and physical communities, xii, 204–5
  three domains of, 204–5
  utopian and dystopian versions of,
    7, 18–19n, 52, 57, 74–5, 83, 98, 115,
    120n, 128, 157, 179, 189, 191–2,
    194–5, 202, 203, 224n
  and virtual *flânerie*, 199–201, 210, 223–4n
virtual reality, 16, 72, 94, 114, 121n, 130,
    135, 190, 196
  and cyberspace, ix, 44–6, 49, 122, 201
'virtual urbanization' perspective, 67–9, 91
'virtuvoltage', 46
virus, 105
Voog, Anna, 220–1

Walt Disney Studios, 120
Wark, McKenzie, 38, 97–8, 223n
Wasserman, I.M., 214
Watergate, 165n, 208
Watts, Duncan, 189
Weaver, W., 55–6, 82n
Wehner, Joseph, 76–77, 79, 81
the WELL, 47, 61–2, 132, 225n

Wellman, Barry, 17n, 194, 202, 204
Wenger, E., 176
Wertheim, Margaret, 190–1
Whitaker, R., 223n
White, Mimi, 1–3, 17n
Whittle, D., 195, 224n
Wiener, Norbert, 56–57
Williams, Christopher, 18n
Williams, Raymond, 85–6, 91–2, 120n, 140,
    161–2, 178–9
  mobile privatization, xii, 88, 91–3, 98,
    99–100, 199
  source and agent, 161–2
  technical invention, 140
  technology-as-socially-configured, 140
Willson, Michelle, 195, 225n
Winston, Brian, 13–14, 19n
Wired, 43n, 73, 113, 115, 189, 190, 202
Wise, J. Macgregor, 202
World Bank, protests, 80
World Health Organization (WHO), 222n
World Wide Web (WWW), 47, 50, 60, 75,
    79, 91, 96, 100, 104, 115, 132, 150, 166n,
    189, 212, 216
World-Wide-Wait, 90, 201

Y2K bug, 2

Zettl, H., 106